Notes on Numerical Modeling in Geomechanics

T0236600

Notes on Numerical Modeling in Geomechanics

William G. Pariseau

Professor Emeritus, Department of Mining Engineering,
University of Utah, Salt Lake City, USA

CRC Press
Taylor & Francis Group
Boca Raton London New York Leiden

CRC Press is an imprint of the
Taylor & Francis Group, an **informa** business

A BALKEMA BOOK

Cover: Twin arched back tunnels 6.7 m (22 ft) wide; width-to-height ratio is one; pillar width is also 6.7 m (22 ft). Colors represent different rock formations.

First published 2022
by CRC Press/Balkema

Schipholweg 107C, 2316 XC Leiden, The Netherlands
e-mail: enquiries@taylorandfrancis.com
www.routledge.com – www.taylorandfrancis.com

CRC Press/Balkema is an imprint of the Taylor & Francis Group, an informa business

Library of Congress Cataloging-in-Publication Data
A catalog record has been requested for this book

ISBN: 978-0-367-76283-4 (hbk)
ISBN: 978-0-367-76287-2 (pbk)
ISBN: 978-1-003-16628-3 (ebk)

DOI: 10.1201/9781003166283

Typeset in Times New Roman
by codeMantra

Contents

16 Conclusion   93

Preface

These "Notes" are the result of many years of experience in teaching, developing, and practicing numerical modeling in geomechanics; they are intended for beginners with a background in differential equations, mechanics of materials, matrix algebra, vector calculus, and some exposure to soil and rock mechanics.

The primary objective of these "Notes" is to develop a basic understanding of the finite element method (FEM) as used in geomechanics. FEM is well understood and has been in undergraduate engineering curricula for many years; it is by far the most popular numerical method for engineering design. Although the best way to learn FEM is by doing, that is, by writing a finite element computer program, learning a computer programming language such as Fortran or some version of C or another high-level language is not the undergraduate engineering requirement that it once was. Consequently, an approach based on development of FEM concepts and then use of available FEM programs to demonstrate applications is followed. A similar approach is followed in the case of the boundary element method (BEM) and the distinct element method (DEM).

However, programming comments are given at the end of each chapter as additional guidance for those who are familiar with a high-level programming language such as Fortran. Otherwise, these comments are easily skipped in the progress of the "Notes".

After a brief introduction in Chapter 1, an intuitive approach to the concept of a finite element is applied to interpolation over a triangle in Chapter 2. Derivatives of interpolation functions are discussed in Chapter 3. Linear interpolation over a quadrilateral and derivatives are discussed in Chapters 4 and 5, respectively. Element equilibrium and stiffness are developed in Chapter 6, followed by development of global equilibrium and stiffness in Chapter 7. Chapter 8 describes the concept of static condensation and the four-element constant-strain quadrilateral. Equation solving by elimination and iteration schemes is discussed in Chapter 9. Material nonlinearity and time integration are developed in Chapters 10 and 11. Fundamentals of seepage analysis and the finite element approach to hydro-mechanical coupling are described in Chapters 12 and 13, respectively. Basics of BEM are described in Chapter 14; DEM basics are described in Chapter 15.

Short illustrative example computations and longer more detailed studies from engineering practice are presented at various stages of model developments.

Problems and questions for home study are given in an appendix and are paced for a 15-week semester course; solutions are also provided.. A summary is given in Chapter 16, followed by a list of references consulted in the writing of these "Notes".

Acknowledgments

Numerical modeling in geomechanics is a well-developed discipline as evident in the extensive literature on the subject and the fact that the subject has been offered at the undergraduate level of study for many years. This text draws from that literature, from programming the finite element method from the early beginnings in the 1960s, and from teaching the finite element method to undergraduates and graduate students since. Grateful acknowledgment is made to those students for their patience and perseverance despite the lack of a suitable textbook that combines numerical analysis with geomechanics complete with example problems and problems for home study.

Illustrations are essential to learning, and grateful acknowledgment is expressed for financial assistance from the Department of Mining Engineering, University of Utah, for rendering figures suitable for publication. In this regard, use of figures from the following sources is most gratefully acknowledged:

- Cambridge University Press.
- Department of the Army, Missouri River Division, Corps of Engineers.
- Taylor & Francis, CRC Press.
- Amit Acharya, *A Discrete Element Approach to Ball Mill Mechanics*, M.S. Thesis, University of Utah, 1991.
- Chang-Ha Ryu, *Numerical Modeling of Fragmentation in Jointed Geologic Media*, Ph.D. Dissertation, University of Utah, 1989.
- Gen-Hua Shi, *Discontinuous Deformation Analysis: A New Numerical Model for the Statics and Dynamics of Block Systems*, Ph.D. Dissertation, University of California, Berkeley, 1988.

About the author

William G. Pariseau

Professor Pariseau obtained his B.S. degree in Mining Engineering at the University of Washington (Seattle) following the geological option and subsequently earned a Ph.D. in Mining Engineering at the University of Minnesota with an emphasis on rock mechanics and with a minor in applied mathematics. Prior to his Ph.D., he obtained practical experience working for the City of Anchorage, the Alaska Department of Highways, the Mineral Resources Division of the U.S. Bureau of Mines (Spokane), the Anaconda Copper Co. in Butte, Montana, and the New York-Alaska Gold Dredging Corp. in Nyac, Alaska. He served in the United States Marine Corps (1953–1956). He maintained a strong association with the former U.S. Bureau of Mines, first with the Pittsburgh Mining Research Center and later with the Spokane Mining Research Center. He continues his association with the now Spokane and Pittsburg Mining Research Divisions of the National Institute of Occupational Health and Safety. He is a registered professional engineer and has consulted for a number of commercial and government entities.

Currently, he is a professor *emeritus* and former holder of the Malcolm McKinnon endowed chair in mining engineering at the University of Utah. He joined the department in 1971 following academic appointments at the Montana College of Science and Technology and the Pennsylvania State University. He has been a visiting academic at Brown University, Imperial College London and the Commonwealth Science and Industrial Research Organization (CSIRO), Australia. He and his colleagues have received a number of rock mechanics awards; he was recognized as a distinguished university research professor at the University of Utah in 1991. In 2010, he was recognized for teaching in the College of Mines and Earth Sciences with the Outstanding Faculty Teaching Award. In the same year, he was honored by the Old Timers Club with their prestigious Educator Award. He was honored as a Fellow of the American Rock Mechanics Association in 2015.

Chapter 1

Introduction

Interest here is mainly in developing sufficient understanding of the finite element method (FEM) for intelligent application and less in the purely mathematical features of the technique. The approach is to learn first through development of basic concepts that collectively form the core of most finite element packages and then to put these concepts into practice using available commercial codes for solution of example problems of interest.

A review of fundamentals of stress, strain, and stress–strain relations in linear elasticity is offered in an appendix which is more detailed than what is usually presented in a first course in mechanics of materials, but may prove useful. Although presented in an appendix, an initial review of fundamentals of stress and strain is usually helpful before proceeding to numerical modeling.

Development of basic numerical modeling concepts begins with an intuitive approach to interpolation over a triangular element. The vertices or corners of an element are "nodes" in finite element jargon. An element is considered to be a homogeneous subdivision of a material body, and the simplest shape for covering a two-dimensional body is a triangle. A tetrahedron is an example of a three-dimensional finite element. An assembly of such elements may be used to approximate the geometry of a region in three dimensions. Other element shapes are certainly possible, for example quadrilaterals and brick-shaped elements.

Interpolation on a triangle is considered first. Matrix notation is used to compact notation. For example, interpolation of temperature $T(x, y)$ over a triangle leads to $T(x, y) = [N(x, y)]\{T_n\}$, where $\{T_n\}$ is a 3×1 column matrix of known temperatures at the three nodes of the considered triangle ($n = 3$), $[N(x, y)]$ is a 1×3 matrix of known interpolation functions $N(x, y)$, and $T(x, y)$ is the unknown temperature at any point on the triangle including the edges and vertices (nodes). The notation $[N]$ is a much used finite element technology.

Computer programs are modular and consist of subprograms or modules for performing specific tasks such as computation of an interpolation matrix. A node displacement to element strain module readily follows. For example, $\{\varepsilon\} = [B]\{\delta\}$, where $\{\varepsilon\}$, $[B]$, and $\{\delta\}$ are a 3×1 matrix of strains, a 3×6 matrix of a derivatives of $[N]$, and a 6×1 column matrix of x and y displacements u and v at the nodes of the considered triangle. Elaboration to quadrilaterals is done next, followed by an easy two-dimensional stress–strain module: $\{\sigma\} = [C]\{\varepsilon\}$, where $\{\sigma\}$, $[C]$, and $\{\varepsilon\}$ are a 3×1 matrix of stress, a 3×3 matrix of elastic constants, and a 3×1 matrix of strains, respectively.

DOI: 10.1201/9781003166283-1

Further progress depends on the famous divergence theorem that allows transformation between surface and volume integrals and provides for the introduction of the principle of virtual work. An element stiffness matrix $[k]$ follows, which defines and links node forces $\{f\}$ to node displacements $\{\delta\}$. Thus, $\{f\} = [k]\{\delta\}$. Numerical integration is then discussed within the context of computing element stiffness matrices.

Once approximation within an element is achieved, assembly of the array of elements, representing the body of interest, must be considered in accordance with basic physical laws, kinematics, and material laws. Mass must be conserved; requirements for equilibrium, as dictated by balances of linear and angular momentum, must be met; the geometry of motion relating strains to displacements must be compatible and relationships between stress and strain are observed, all are compactly stated with the aid of the well-known divergence theorem. Requirements for global equilibrium and formulation of a master stiffness matrix are then done. Thus, $\{F\} = [K]\{\Delta\}$, where $\{F\}, [K]$, and $\{\Delta\}$ are a column matrix of global forces, a global stiffness matrix, and a column matrix of global displacements, respectively.

A digression to development of a quadrilateral formed by four constant-strain triangles leads to consideration of static condensation and then to numerical equation solving. Gauss elimination and iteration schemes are developed. Additional topics briefly discussed include material nonlinearity and time integration.

Pressures in fluid flow through porous media may be approximated in the same manner as temperature. Similarly, pressure gradients may be obtained directly from pressures by differentiation. Solid stress and fluid velocity are then obtained through Hooke's law in elastic analyses and Darcy's law in seepage analyses. The essentials of seepage analysis are presented. Discussions of BEM, DEM, and coupled fluid flow and solid deformation follow.

Some comments about programming or writing code for a finite element program may be helpful and provide some insights into structures of commercial codes. Essentially, a mainline program serves as an executive routine that calls other subprograms to do specialized tasks during a program run. A mainline program is needed to coordinate the specialized tasks assigned to suggested subroutines in a finite element program. How much computation to do in a particular subroutine is generally a matter of taste. However, most finite element programs share these tasks:

(1) Reading input data......................................	RDRN
(2) Interpolation matrix computation........................	NMTX
(3) Strain–displacement matrix computation.................	BMTX
(4) Material properties matrix formulation.................	EMTX
(5) Element stiffness computation........................	KMTX
(6) Assembly of the global stiffness matrix...............	SMTX
(7) Insertion of displacement boundary conditions..........	DBCD
(8) Equation solving.................................	SLVR
(9) Element stresses and strains computation...............	SGPS
(10) Writing output data..................................	WRTR

The names on the right-hand side are for identification of subprograms or modules with MTX meaning "matrix".

Input data files that establish the region of interest are mainly: (1) a file with specification of elements and associated corner points and (2) a file with specification of coordinates of all nodes. Together, these two files are the finite element "mesh". They tend to be large files and almost always need to be generated automatically by a preprocessing package or mesh generator. Generation manually using a large board digitizer is possible, but challenging. Mesh generation even automatically is by far the most laborious and most important task in finite element analysis. In geomechanics, mesh generation is especially difficult because of the presence of different rock types, sinuous geologic contacts, faults, and complex excavation geometries and sequences. Topography is yet another complication in geomechanics analyses.

Displacement and force boundary conditions may be part of the mesh files or included in a separate and much smaller file. Material properties are contained in a rather small input file required in establishing a program run stream (a list of files and other data pertinent to analysis). A preexcavation stress field is generally needed in the form of a stress state specification for each element. Although this initial stress field file is large, it is also simple to specify as part of the input data files. Some finite element programs allow for element excavation and require a small file containing a list of elements to be excavated. Some finite element programs create an excavation in advance by surrounding an excavation with elements while leaving vacant the excavation region proper. In either case, excavation geometry is a specified input file. Additionally, a list of control numbers such as the number of elements in the mesh, the number of nodes, output information and so forth are usually needed for input

Thus, a run stream for a finite element analysis contains input files:

(1) elements defined by corner points (nodes)
(2) node coordinates
(3) displacement and force boundary conditions
(4) material properties
(5) preexcavation stress field
(6) excavation geometry
(7) excavation sequence
(8) control numbers.

Output information from a finite element analysis can be voluminous and varied, but usually includes files of node displacements, element strains, and element stresses. Understanding the results of an analysis of stress and obtaining guidance from output data are not easy tasks. Output files may easily be several million lines long. For this reason, automatic post-processing of finite element data and indeed data from numerical analyses packages in general are almost always needed for proper interpretation. In this regard, color graphics are the usual way that output data are presented to a program user. Of course, there is great variety in such presentations.

REFERENCES

Bathe K.-J. (1982) *Finite Element Procedures in Engineering Analysis.* Prentice-Hall, Inc., Englewood Cliffs, NJ, pp 735.

Cook, R.D. (1974) *Concepts and Applications of Finite Element Analysis.* John Wiley & Sons, Inc., New York, pp 402.

Cook, R.D., D.S. Malkus and M.E. Plesha (1989) *Concepts and Applications of Finite Element Analysis* (3rd ed). John Wiley & Sons, Inc., New York, pp 630.

Desai, C.S. and J.F. Abel (1972) *Introduction to the Finite Element Method.* Van Nostrand Reinhold Company, New York, pp 477.

Fenner, R.T. (1972) *Finite Element Methods for Engineers.* Macmillan Press, London, pp 171.

Lawn, B.R. and T.R. Wilshaw (1975) *Fracture of Brittle Solids.* Cambridge University Press, Cambridge, pp 204.

Logan, D.L. (1986) *A First Course in the Finite Element Method.* PWS-Kent Publishing Company, Boston, MA, pp 617.

Logan, D.L. (1993) *A First Course in the Finite Element Method* (2nd ed). PWS Publishing Company, Boston, MA, pp 662.

Oden, J.T. (1972) *Finite Elements of Nonlinear Continua.* McGraw-Hill, New York, pp 432.

Przemieniecki, J. S. (1968) *Theory of Matrix Structural Analysis.* McGraw-Hill, New York, pp 468.

Segerlind, L.J. (1976) *Applied Finite Element Analysis.* John Wiley & Sons, Inc., New York, pp 422.

Zienkiewicz, O.C. and Y.K. Cheung (1967) *The Finite Element Method in Structural and Continuum Mechanics.* McGraw-Hill, New York. pp 272.

Zienkiewicz, O.C. (1977) *The Finite Element Method* (3rd ed). McGraw-Hill, New York, pp 787.

As one might surmise from the publishing dates, these books were consulted early in the development of the subject. The book by Cook was most helpful and is now expanded into a third edition. The book by Desai is oriented toward geomechanics in contrast to the structural orientation of the others and includes seepage analysis and has been especially helpful for this reason. Finite element analysis first appeared in the mid-1950s, with the advent of the computer; it is now a mature subject with an enormous literature that includes many other worthwhile texts, reference books, conference proceedings, and journal articles.

Chapter 2

Interpolation over a triangle

The triangle is a simple, but very useful shape for developing a basic understanding of the fundamentals of finite element analysis of stress and seepage in two dimensions. The unknowns in stress analyses are almost always the displacements. In seepage analysis, the unknowns are usually fluid pressures. Once an adequate theory is developed for interpolating the unknowns in the interior of a triangle from the known values at the triangle corners, explicit expressions for linear interpolation functions can be obtained that are suitable for programming into a finite element package. However, the numerical quality of a linear triangle is low and is used here only to illustrate the concept of interpolation in a simple intuitive manner.

2.1 LINEAR THEORY

Consider a problem of approximating a two-dimensional displacement field in the interior of a triangular area. Components of displacement relative to the x, y system of coordinates chosen for reference are u and v, respectively, as shown in Figure 2.1. A very crude approximation would be to suppose that u and v are constant in the triangle. Derivatives of constants are zero, of course, so if strains are of interest, then such a crude approximation would not be useful.

The next step in this intuitive approach to the problem is to suppose that the displacements vary linearly with x and y. Thus,

$$u = a_0 + a_1 x + a_2 y$$
$$v = b_0 + b_1 x + b_2 y$$

(2.1)

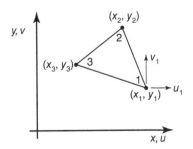

Figure 2.1 Notation for a triangular element in stress analyses.

DOI: 10.1201/9781003166283-2

If the approximation Equation (2.1) is to be useful in actual calculations, the constants $a_0, a_1, ..., b_2$ must be known. In fact, they can be determined in terms of the displacements $u_1, u_2, ..., v_3$ at the corners of the triangle. Substituting the corner coordinates $x_1, y_1, ..., x_3, y_3$ into (2.1) gives

$$
\begin{aligned}
u_1 &= a_0 + a_1 x_1 + a_2 y_1 \\
u_2 &= b_0 + b_1 x_2 + b_2 y_2 \\
u_3 &= b_0 + b_1 x_3 + b_2 y_3
\end{aligned}
\tag{2.2}
$$

which in matrix notation is

$$
\begin{Bmatrix} u_1 \\ u_2 \\ u_3 \end{Bmatrix} =
\begin{bmatrix}
1 & x_1 & y_1 \\
1 & x_2 & y_2 \\
1 & x_3 & y_3
\end{bmatrix}
\begin{Bmatrix} a_0 \\ a_1 \\ a_2 \end{Bmatrix}
\tag{2.3}
$$

or in more compact form is

$$
\{u_n\} = [A(x_m, y_m)]\{a\}
\tag{2.4}
$$

where the curly braces {} denote a column matrix and the brackets [] denote a rectangular matrix. This system of three equations in three unknowns can be solved in several ways, for example, by the method of determinants. Symbolically, the solution is

$$
\{a\} = [A]^{-1}\{u_n\}
\tag{2.5}
$$

where $[A]^{-1}$ is the inverse of $[A]$. Back-substitution gives the form

$$
(u)_{(1\times1)} = \{1 \ x \ y\}_{(1\times3)}[A]^{-1}_{(3\times3)}\{u_n\}_{(3\times1)}
\tag{2.6}
$$

where the dimensions of the various matrices are given in the subscripts enclosed in parentheses. Note that a scalar is a 1×1 matrix here and the curly brackets {} denote column and row matrices in Equation (2.6).

Multiplying the first two matrices on the right-hand side of Equation (2.5) leads to the form

$$
(u) = \{N(x,y)_1 \ N(x,y)_2 \ N(x,y)_3\}_{(1\times3)}\{u_n\}_{(3\times1)}
\tag{2.7}
$$

where the dependency of Equation (2.7) on x and y is emphasized. A similar derivation leads to an almost identical expression for v. Thus,

$$
(v) = \{N(x,y)_1 \ N(x,y)_2 \ N(x,y)_3\}_{(1\times3)}\{v_n\}_{(3\times1)}
\tag{2.8}
$$

Equations (2.7) and (2.8) in extended form are

$$
\begin{aligned}
u &= N_1 u_1 + N_2 u_2 + N_3 u_3 \\
v &= N_1 v_1 + N_2 v_2 + N_3 v_3
\end{aligned}
\tag{2.9}
$$

where the dependency on x and y is implicit.

In the case of temperature T and fluid pressure p, approximation, that is interpolation, to an interior point, is done in much the same manner. Thus,

$$
\begin{aligned}
T &= N_1 T_1 + N_2 T_2 + N_3 T_3 \\
p &= N_1 p_1 + N_2 p_2 + N_3 p_3
\end{aligned}
\tag{2.10}
$$

Any other quantity of interest could be approximated over a triangle in the same way. To be sure, the functional coefficients N_1, N_2, N_3 are the same functions as before, that is, interpolation functions.

Equations (2.9) can be expressed in more compact form:

$$
\begin{Bmatrix} u \\ v \end{Bmatrix}_{(2\times1)} = [N]_{(2\times6)} \{\delta\}_{(6\times1)}
\tag{2.11}
$$

where the N-matrix has the form

$$
[N] = \begin{bmatrix}
N_1 & N_2 & N_3 & 0 & 0 & 0 \\
0 & 0 & 0 & N_1 & N_2 & N_3
\end{bmatrix}
\tag{2.12}
$$

provided the column matrix of node displacements $\{\delta\}$ is organized by components:

$$
\{\delta\} = \begin{Bmatrix} u_1 \\ u_2 \\ u_3 \\ v_1 \\ v_2 \\ v_3 \end{Bmatrix}
\tag{2.13}
$$

An alternative and commonly used organization of the node displacement matrix is by nodes:

$$
\{\delta\} = \begin{Bmatrix} u_1 \\ v_1 \\ u_2 \\ v_2 \\ u_3 \\ v_3 \end{Bmatrix}
\tag{2.14}
$$

in which case

$$
[N] = \begin{bmatrix}
N_1 & 0 & N_2 & 0 & N_3 & 0 \\
0 & N_1 & 0 & N_2 & 0 & N_3
\end{bmatrix}
\tag{2.15}
$$

2.2 EXPLICIT FORMULAS

Programming requires explicit expressions for each of the three interpolation functions, Equation (2.12), whether the problem involves displacements, temperature, or pressure. The solution to Equations (2.2) provides a guidance in the linear case. Thus,

$$
a_0 = \begin{vmatrix} u_1 & x_1 & y_1 \\ u_2 & x_2 & y_2 \\ u_3 & x_3 & y_3 \end{vmatrix} \Big/ \Delta
\tag{2.16}
$$

$$
a_1 = \begin{vmatrix} 1 & u_1 & y_1 \\ 1 & u_2 & y_2 \\ 1 & u_3 & y_3 \end{vmatrix} \Big/ \Delta
\tag{2.17}
$$

$$
a_2 = \begin{vmatrix} 1 & x_1 & u_1 \\ 1 & x_2 & u_2 \\ 1 & x_3 & u_3 \end{vmatrix} \Big/ \Delta
\tag{2.18}
$$

$$
\Delta = \begin{vmatrix} 1 & x_1 & y_1 \\ 1 & x_2 & y_2 \\ 1 & x_3 & y_3 \end{vmatrix}
\tag{2.19}
$$

where the vertical bars $|\,|$ denote determinants and Δ is the determinant of the coefficients in Equations (2.2). After some algebra and back-substitution of the expanded forms of Equations (2.16–2.19) into Equation (2.1a), one obtains

$$
\begin{aligned}
u = \;& (P_1 x + P_4 y + P_7)(1/\Delta)u_1 \\
& + (P_2 x + P_5 y + P_8)(1/\Delta)u_2 \\
& + (P_3 x + P_6 y + P_9)(1/\Delta)u_3
\end{aligned}
\tag{2.20}
$$

where

$$
\begin{aligned}
P_1 &= y_2 - y_3, & P_4 &= x_3 - x_2, & P_7 &= (x_2 y_3 - x_3 y_2) \\
P_2 &= y_3 - y_1, & P_5 &= x_1 - x_3, & P_8 &= (x_3 y_1 - x_1 y_3) \\
P_3 &= y_1 - y_2, & P_6 &= x_2 - x_1, & P_9 &= (x_1 y_2 - x_2 y_1)
\end{aligned}
\tag{2.21}
$$

Thus,

$$
\begin{aligned}
N_1 &= (P_1 x + P_4 y + P_7)/\Delta \\
N_2 &= (P_2 x + P_5 y + P_8)/\Delta \\
N_3 &= (P_3 x + P_6 y + P_9)/\Delta
\end{aligned}
\tag{2.22}
$$

Consideration of algebraic details shows that $\Delta = P_1 + P_2 + P_3$ and also that the area A of a triangle is $\Delta/2$. The linear interpolation functions in Equations (2.22) are therefore determined by the nine numbers given in Equations (2.21).

An interesting feature of these interpolation functions is that at the 1-corner where the coordinates are (x_1, y_1), $N_1 = 1$. At the 2- and 3-corners, $N_1 = 0$. Similar relations hold at the other two corners of the triangle. These results can be inferred from Equations (2.20) by simply substituting corner coordinates. For example, at (x_1, y_1), one must have $u = u_1$ and so forth.

The interpolation functions Equations (2.22) have potential applications outside finite analysis. An important example is interpolation of data from an irregular array of points connected by triangles to a regular array connected by rectangles, that is, to a regular grid illustrated in Figure 2.2. Almost all packaged contouring routines require data distributed over a rectangular grid. However, much natural data are acquired at points not nearly so conveniently spaced. Interpolation to a regular grid is then necessary prior to contouring.

Computation of the interpolation functions begins by translating global coordinates of the corners of a triangle to local coordinates. If the global node numbers of the shaded triangle in Figure 2.2 are 15, 22, and 37 corresponding to local corner numbers 1, 2, and 3, then in subscript notation:

$$x_1 = xord_{15}, \ \ y_1 = yord_{15}$$
$$x_2 = xord_{22}, \ \ y_2 = yord_{22} \tag{2.23a}$$
$$x_3 = xord_{37}, \ \ y_3 = yord_{37}$$

In another notation achieved by simply moving subscripts to following parentheses:

$$x(1) \ = \ xord(15), \ \ y(1) = yord(15)$$
$$x(2) \ = \ xord(22), \ \ y(2) = yord(22) \tag{2.23b}$$
$$x(3) \ = \ xord(37), \ \ y(3) = yord(37)$$

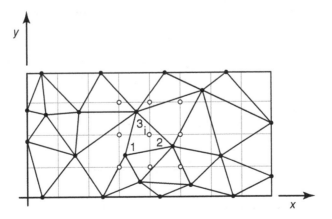

Figure 2.2 Data interpolation from an irregular triangular network to a rectangular grid.

where the arrays *xord*(*n*) and *yord*(*n*) are global coordinates of the *n*th node (*n* = 15, 22, and 37 in this example) and *x*(*m*), *y*(*m*), *m* = 1, 2, 3, are the corresponding local node (corner) coordinates. In fact, the coordinate values are the same; only the node numbers change from global to local designations. Global numbering of nodes may proceed in any sequence. However, local numbering associated with each triangular element should proceed in a counter-clockwise direction (assuming right-handed co-ordinate systems). If the direction is clockwise, the sign of the determinant, Δ, will be reversed and the associated area A will be negative.

2.3 LINEAR STRAIN TRIANGLES

Linear interpolation of displacements over a triangle leads to constant strains within the triangle and jumps in strains between adjacent triangles. A smoother strain ap-proximation can be obtained by a higher-order displacement approximation. The next increase in order of approximation involves squares and products of the coordinates and a considerable increase in complexity. Indeed, the analysis shows that explicit formulas are impractical, unlike the linear displacement approximation.

Consider the displacement in the *x*-direction interpolated by a second-order polynomial containing all terms. Thus,

$$u = a_0 + a_1 x + a_2 y + a_3 x^2 + a_4 xy + a_5 y^2 \qquad (2.24)$$

There are six unknown coefficients in Equation (2.24), so six known values of *u* are needed. The three corner displacements, u_1, u_2, u_3, are available for this purpose. An additional three values, u_4, u_5, u_6, can be obtained from the three mid-side points shown in Figure 2.3. The choice of the mid-side points is intuitively reasonable in consideration of alternative choices.

Substitution of the known node point coordinates and displacements into Equation (2.24) gives a system for the six unknown coefficients in Equation (2.24). Thus,

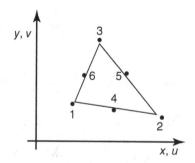

Figure 2.3 A quadratic displacement, linear strain triangle.

$$\begin{Bmatrix} u_1 \\ u_2 \\ u_3 \\ u_4 \\ u_5 \\ u_6 \end{Bmatrix} = \begin{bmatrix} 1 & x_1 & y_1 & x_1^2 & x_1 y_1 & y_1^2 \\ 1 & x_1 & y_1 & x_1^2 & x_1 y_1 & y_1^2 \\ 1 & x_2 & y_2 & x_2^2 & x_2 y_2 & y_2^2 \\ 1 & x_3 & y_3 & x_3^2 & x_3 y_3 & y_3^2 \\ 1 & x_4 & y_4 & x_4^2 & x_4 y_4 & y_4^2 \\ 1 & x_5 & y_5 & x_5^2 & x_5 y_5 & y_5^2 \\ 1 & x_6 & y_6 & x_6^2 & x_6 y_6 & y_6^2 \end{bmatrix} \begin{Bmatrix} a_o \\ a_1 \\ a_2 \\ a_3 \\ a_4 \\ a_5 \\ a_6 \end{Bmatrix} \qquad (2.24a)$$

which in compact form is

$$\{u_n\}_{(6\times1)} = [A']_{(6\times6)} \{a\}_{6\times1} \qquad (2.24b)$$

After inversion,

$$\{a\}_{(6\times1)} = [A]_{(6\times6)} \{u_n\}_{6\times1} \qquad (2.24c)$$

where $[A]$ is the inverse of $[A']$. Back-substitution into Equation (2.24) gives

$$\begin{aligned}
u = \; & (A_{11}u_1 + A_{12}u_2 + A_{13}u_3 + A_{14}u_4 + A_{15}u_5 + A_{16}u_6) \\
& + x(A_{21}u_1 + A_{22}u_2 + A_{23}u_3 + A_{24}u_4 + A_{25}u_5 + A_{26}u_6) \\
& + y(A_{31}u_1 + A_{32}u_2 + A_{33}u_3 + A_{34}u_4 + A_{35}u_5 + A_{36}u_6) \\
& + x^2(A_{41}u_1 + A_{42}u_2 + A_{43}u_3 + A_{44}u_4 + A_{45}u_5 + A_{46}u_6) \\
& + xy(A_{51}u_1 + A_{52}u_2 + A_{53}u_3 + A_{54}u_4 + A_{55}u_5 + A_{56}u_6) \\
& + y^2(A_{61}u_1 + A_{62}u_2 + A_{63}u_3 + A_{64}u_4 + A_{65}u_5 + A_{66}u_6)
\end{aligned} \qquad (2.24d)$$

Inspection of Equation (2.24c) shows that there are now 36 coefficients (as) to be computed. Equation (2.24d) can be rearranged into a standard form. Thus,

$$u = N_1 u_1 + N_2 u_2 + N_3 u_3 + N_4 u_4 + N_5 u_5 + N_6 u_6 \qquad (2.24e)$$

The six N interpolation functions in Equation (2.24d) depend on (x, y, x^2, xy, y^2). An explicit formula for these coefficients and interpolation functions is possible in principle, but would require expansion of the inverse matrix $[A']$ which would be pages long. As a practical matter, numerical inversion is required.

Differentiation of Equation (2.23d) with respect to x gives the normal strain ε_{xx}. Thus,

$$\begin{aligned}
\varepsilon_{xx} = \frac{\partial u}{\partial x} = \; & (A_{21}u_1 + A_{22}u_2 + \ldots + A_{26}u_6) \\
& + 2x(A_{41}u_1 + A_{42}u_2 + \ldots + A_{46}u_6) \\
& + y(A_{51}u_1 + A_{52}u_2 + \ldots + A_{56}u_6)
\end{aligned} \qquad (2.24f)$$

```
        Do 200 j=1,3
        n=NOD(nel,j)
        x(j)=xord(n)
200     y(j)=yord(n)
        P(7)=(x(2)*y(3)-x(3)*y(2))
        P(8) =(x(3)*y(1)-x(1)*y(3))
        P(9)=(x(1)*y(2)-x(2)*y(1))
        delta=p(7)+p(8)+p(9)
c
        P(1) = ( y(2)-y(3) )/delta
        P(2) = ( y(3)-y(1) )/delta
        P(3) = ( y(1)-y(2) )/delta
        P(4) = ( x(3)-x(2) )/delta
        P(5) = ( x(1)-x(3) )/delta
        P(6) = ( x(2)-x(1) )/delta
c
        P(7)=P(7)/delta
        P(8)=P(8)/delta
        P(9)=P(9)/delta

        A = 0.5*delta
```

Figure 2.4 Statements for computing interpolation function coefficients.

Although the development path is now evident, at least for the triangle, one may surmise that quadratic displacement (linear strain) and even higher-order elements require much more effort than linear displacement (constant-strain) elements. In this regard, trade-offs between computation time and data storage capacity and handling time need to be considered, especially when nonlinear material behavior is anticipated. Much depends on available computer facilities, of course.

2.4 PROGRAMMING COMMENTS

The correspondence between local and global node numbers must be specified for each triangle in Figure 2.2. A two-dimensional array NOD(i, j) serves this purpose. In this array, i is the triangle number. If the number of triangles is *nelem*, then $i = 1$, 2,..., *nelem*. The index $j = 1$, 2, 3, so $n_1 = $ NOD(i, 1) is the global node number of the first corner of the ith triangle. In this example, $n_1 = 15$. Similarly, $n_2 = 22 = $ NOD(i, 2) and $n_3 = 37 = $ NOD(i, 3). The Fortran statements in Figure 2.4 calculate the coefficients required for interpolation within element *nel*. If "delta" is zero or negative, an error occurs, so an if-statement trap for detecting non-positive delta would be a wise addition following the computation of delta. Area A is computed for later use.

Chapter 3

Derivatives of interpolation functions

Derivatives of the displacement interpolation functions are needed in the analysis of stress for the computation of strains and in the analysis of seepage for the computation of the hydraulic (pressure) gradient. When linear interpolation is used, the derivatives are constant. Hence, linear displacement triangles are also constant-strain triangles. They are also constant-stress triangles in elastic analysis because of the linear relationship between stress and strain (Hooke's law). In seepages analysis, the elements are constant hydraulic gradient elements and also constant fluid velocity elements because of the linear relationship between hydraulic gradient and fluid velocity (Darcy's law). The finite element form of interpolation function derivatives follows directly from the definitions of strain and hydraulic gradient.

3.1 STRAINS

The strain–displacement relations for small strains relative to rectangular coordinates are

$$
\begin{aligned}
\varepsilon_{xx} &= \frac{\partial u}{\partial x}, \quad \gamma_{xy} = \frac{\partial u}{\partial y} + \frac{\partial v}{\partial x} \\
\varepsilon_{yy} &= \frac{\partial v}{\partial y}, \quad \gamma_{yz} = \frac{\partial v}{\partial z} + \frac{\partial w}{\partial y} \\
\varepsilon_{zz} &= \frac{\partial w}{\partial z}, \quad \gamma_{zx} = \frac{\partial w}{\partial x} + \frac{\partial u}{\partial z}
\end{aligned}
\tag{3.1}
$$

where ε is a normal strain, γ is an engineering shear strain, and u, v, and w are displacement components in the x-, y-, and z-directions, respectively.

In either plane strain or plane stress analyses, the in-plane x–y strains are computed from the displacements according to Equations (3.1). In consideration of the linear displacement field used in interpolation over the 3-node triangle, the strains are simply

$$
\varepsilon_{xx} = a_1, \quad \varepsilon_{yy} = b_2 \quad \gamma_{xy} = \gamma_{yx} = a_2 + b_1
\tag{3.2}
$$

DOI: 10.1201/9781003166283-3

In terms of the node displacements,

$$
\begin{aligned}
\varepsilon_{xx} &= [P_1 u_1 + P_2 u_2 + P_3 u_3]/\Delta \\
\varepsilon_{xx} &= [P_4 v_1 + P_5 v_2 + P_6 v_3]/\Delta \\
\gamma_{xy} &= [P_4 u_1 + P_5 u_2 + P_6 u_3]/\Delta \\
&\quad + [P_1 v_1 + P_2 v_2 + P_3 v_3]/\Delta
\end{aligned}
\tag{3.3}
$$

In matrix form, Equations (3.3) are

$$
\begin{Bmatrix} \varepsilon_{xx} \\ \varepsilon_{yy} \\ \gamma_{xy} \end{Bmatrix} = (1/\Delta)
\begin{bmatrix}
P_1 & P_2 & P_3 & 0 & 0 & 0 \\
0 & 0 & 0 & P_4 & P_5 & P_6 \\
P_4 & P_5 & P_6 & P_1 & P_2 & P_3
\end{bmatrix}
\begin{Bmatrix} u_1 \\ u_2 \\ u_3 \\ v_1 \\ v_2 \\ v_3 \end{Bmatrix}
\tag{3.4}
$$

where the node displacements are organized by components. A common alternative is

$$
\begin{Bmatrix} \varepsilon_{xx} \\ \varepsilon_{yy} \\ \gamma_{xy} \end{Bmatrix} = (1/\Delta)
\begin{bmatrix}
P_1 & 0 & P_2 & 0 & P_3 & 0 \\
0 & P_4 & 0 & P_5 & 0 & P_6 \\
P_4 & P_1 & P_5 & P_2 & P_6 & P_3
\end{bmatrix}
\begin{Bmatrix} u_1 \\ u_2 \\ u_3 \\ v_1 \\ v_2 \\ v_3 \end{Bmatrix}
\tag{3.5}
$$

which is organized by node number. In either case, a common and compact form is

$$
\{\varepsilon\}_{(3\times1)} = [B]_{(3\times6)}\{\delta\}_{(6\times1)}
\tag{3.6}
$$

where the subscripts in parentheses provide a check on the dimensions of the matrices.

Although the displacements $\{\delta\}$ in Equation (3.6) are determined at the nodes, the strains are determined for the element. The B-matrix in Equation (3.6) thus induces a node displacement to element strain transformation. Again, because the displacements vary linearly in the element, the strains as first derivatives of displacement are constants. In an array of such elements, the strains "jump" from element to element. So in linear elastic analysis, the stresses will also be constant within an element and jump from element to element.

3.2 HYDRAULIC GRADIENT

In seepage analysis, the hydraulic gradient is strain-like (derivative of pressure) and may be defined by

$$
h_x = \frac{\partial p}{\partial x} + \gamma_x, \ h_y = \frac{\partial p}{\partial y} + \gamma_y, \ h_z = \frac{\partial p}{\partial z} + \gamma_z
\tag{3.7}
$$

where p is the fluid pressure, say, in MPa, γ is the specific weight of the fluid flowing, and the subscripts indicate components. In this regard, the convention in the technical literature is to align the z-axis with the vertical. However, the xy-plane will be retained here as the analysis plane, so the z-axis is horizontal (when the analysis plane is a vertical section). In view of the linear interpolation scheme adopted for the triangle, the pressure gradients are simply

$$\frac{\partial p}{\partial x} = a_1, \quad \frac{\partial p}{\partial y} = a_2 \tag{3.8}$$

The transformation of node pressure to element pressure gradients is analogous to strains and is

$$\begin{Bmatrix} p_x \\ p_y \end{Bmatrix} = (1/\Delta) \begin{bmatrix} P_1 & P_2 & P_3 \\ P_4 & P_5 & P_6 \end{bmatrix} \begin{Bmatrix} p_1 \\ p_2 \\ p_3 \end{Bmatrix} \tag{3.9}$$

where the subscripts on the pressures on the left-hand side of the equation denote partial derivatives; on the right-hand side, the pressure subscripts indicate node values for the considered triangle. Quite often, pressures are expressed as "heads" using the relation $p = \gamma h$, where h is head (and γ is the specific weight of the fluid); that is, $h = p/\gamma$ and heads are used as the basic unknowns rather than pressures. The only difference is the division by the specific weight of the fluid. A hydraulic potential H is then defined such that the derivatives of H are the head gradients. In this alternative,

$$\begin{Bmatrix} h_x \\ h_y \end{Bmatrix} = (1/\Delta) \begin{bmatrix} P_1 & P_2 & P_3 \\ P_4 & P_5 & P_6 \end{bmatrix} \begin{Bmatrix} H_1 \\ H_2 \\ H_3 \end{Bmatrix} \tag{3.10}$$

where h_x and h_y are partial derivatives of H with respect to x and y. Equation (3.10) in compact form is simply

$$\{h\} = [B_p]\{H\} \tag{3.11}$$

which is analogous to the B-matrix in stress analysis. The subscript p is added to distinguish between the two.

3.3 AXIAL SYMMETRY

Axial symmetry implies no variation with the angular coordinate θ, so all radial planes (rz-planes) are identical. There are strong similarities of axially symmetric analyses to plane analyses, but there are also important differences. In fact, the triangular element under discussion is a ring as shown in Figure 3.1, where the r-displacement component is u and the z-displacement component is w.

The strain–displacement relations in axial symmetry are

$$\varepsilon_{rr} = \partial u/\partial r, \quad \varepsilon_{\theta\theta} = u/r, \quad \varepsilon_{zz} = \partial w/\partial z, \quad \gamma_{rz} = \partial u/\partial z + \partial w/\partial r \tag{3.12}$$

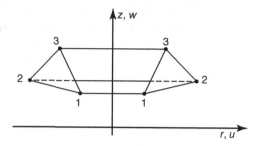

Figure 3.1 Axially symmetric triangular ring element notation.

The displacements are interpolated by

$$
\begin{aligned}
u &= a_0 + a_1 r + a_2 z \\
w &= b_0 + b_1 r + b_2 z
\end{aligned}
\tag{3.13}
$$

Hence, the strains are

$$
\begin{aligned}
\varepsilon_{rr} &= a_1, \ \varepsilon_{\theta\theta} = (a_0 + a_1 r + a_2 z)/r \\
\varepsilon_{zz} &= b_2, \ \gamma_{rz} = a_2 + b_1
\end{aligned}
\tag{3.14}
$$

which in matrix form are

$$
\begin{Bmatrix} \varepsilon_{rr} \\ \varepsilon_{zz} \\ \varepsilon_{\theta\theta} \\ \gamma_{rz} \end{Bmatrix} =
\begin{bmatrix}
P_1 & P_2 & P_3 & 0 & 0 & 0 \\
0 & 0 & 0 & P_4 & P_5 & P_6 \\
P_{10} & P_{11} & P_{12} & 0 & 0 & 0 \\
P_4 & P_5 & P_6 & P_1 & P_2 & P_3
\end{bmatrix}
\begin{Bmatrix} u_1 \\ u_2 \\ u_3 \\ w_1 \\ w_2 \\ w_3 \end{Bmatrix}
\tag{3.15}
$$

where the new coefficients $\left(P_{10} \ P_{11} \ P_2\right)$ are evaluated at the center $\left(r_c, \ z_c\right)$ of the triangle, that is,

$$
\begin{aligned}
P_{10} &= P_7/r_c + P_1 + P_4\left(z_c/r_c\right) \\
P_{11} &= P_8/r_c + P_2 + P_5\left(z_c/r_c\right) \\
P_{12} &= P_9/r_c + P_3 + P_6\left(z_c/r_c\right)
\end{aligned}
\tag{3.16}
$$

In compact form, Equations (3.16) are

$$
\{\varepsilon\}_{(4\times1)} = [B]_{(4\times6)}\{\delta\}_{(6\times1)}
\tag{3.17}
$$

where the node displacement matrix $\{\delta\}$ is organized by components. The option to organize by nodes is also available, of course. With the exception of the third row, programming for initializing the B-matrix in axial symmetry is quite similar to the plane problem B-matrix with the addition of a third row for the calculation of the tangential normal strain which shifts the shear strain down to a fourth row.

In axially symmetric seepage analysis, there is no variable analogous to tangential strain; the B-matrix in Equation (3.11) is still given by Equation (3.10), but with the substitution of r and z for x and y on the left-hand side of the equation. Again, the linear interpolation of pressure in matrix form is

$$(p)_{(1\times1)} = [N(r, z)]_{(1\times3)}\{p_n\}_{(3\times1)} \tag{3.18}$$

If a hydraulic potential H is used instead of pressure, the H may be substituted for p on the left-hand side of Equation (3.18) and H_n for p_n on the right-hand side. The N-matrix and the B-matrix remain unchanged.

3.4 PROGRAMMING COMMENT

Programming for the B-matrix readily follows from the Fortran statements in Figure 2.3. The main task is to initialize the B-matrix for the element of interest. Figure 3.2 shows the Fortran statements for doing this task in accordance with Equation (3.4).

Programming the seepage analysis B-matrix is straightforward and uses the results from Figure 2.3 also. One possible coding is shown in Figure 3.3.

```
        Do 100 m=1,3
        Do 100 n=1,6
100     B(m,n) = 0.0
c
        B(1,1) = P(1)
        B(1,2) = P(2)
        B(1,3) = P(3)
        B(2,4) = P(4)
        B(2,5) = P(5)
        B(2,6) = P(6)
        B(3,1) = P(4)
        B(3,2) = P(5)
        B(3,3) = P(6)
        B(3,4) = P(1)
        B(3,5) = P(2)
        B(3,6) = P(3)
```

Figure 3.2 Initialization of the node displacement to element strain transformation matrix for constant-strain triangles.

```
        Do 100 m=1,2
        Do 100 n=1,3
100     B(m,n) = 0.0
c
        B(1,1) = P(1)
        B(1,2) = P(2)
        B(1,3) = P(3)
        B(2,1) = P(4)
        B(2,2) = P(5)
        B(2,3) = P(6)
```

Figure 3.3 Initialization of the B-matrix in a two-dimensional seepage analysis.

Chapter 4

Linear interpolation for a quadrilateral

Although a triangle is the simplest of all shapes available for approximating a two-dimensional region, a quadrilateral is, in some ways, a more natural shape.[1] For example, a network of quadrilaterals (rectangles) is generated automatically, at least implicitly, with definition of any x–y coordinate system. An increase in analysis effort, including a discussion of a novel coordinate transformation, is required, but the benefits of a quadrilateral element are well worth the effort as demonstrated in subsequent developments.

4.1 THE GENERIC 4-NODE QUADRILATERAL

Consider the generic quadrilateral (square), shown in Figure 4.1, relative to a generic, but rectangular system of coordinates a, b. The interpolation of the a-component of displacement u in generic coordinates is

$$u = \alpha_o + \alpha_1 a + \alpha_2 b + \alpha_3 ab \tag{4.1}$$

where the mixed term ab is added to allow for the unique determination of the four α-coefficients from the known displacements at the four corners of a quadrilateral. Thus,

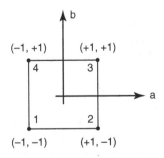

Figure 4.1 A quadrilateral (square in generic coordinates (a, b)).

---

[1] A circle may appear to be even simpler to describe than a triangle. However, an array of circles will not completely fill a two-dimensional region, regardless of any gradation in circle size.

DOI: 10.1201/9781003166283-4

$$
\begin{Bmatrix} u_1 \\ u_2 \\ u_3 \\ u_4 \end{Bmatrix} = \begin{bmatrix} 1 & a_1 & b_1 & a_1b_1 \\ 1 & a_2 & b_2 & a_2b_2 \\ 1 & a_3 & b_3 & a_3b_3 \\ 1 & a_4 & b_4 & a_4b_4 \end{bmatrix} \begin{Bmatrix} \alpha_o \\ \alpha_1 \\ \alpha_2 \\ \alpha_3 \end{Bmatrix} \tag{4.2}
$$

As before, the system of Equation (4.2) may be solved analytically by the method of determinants. Expressions for the coefficients in (1) are thus

$$
\begin{aligned}
\alpha_1 &= (P_1u_1 + P_2u_2 + P_3u_3 + P_4u_4)/\Delta \\
\alpha_2 &= (P_5u_1 + P_6u_2 + P_7u_3 + P_8u_4)/\Delta \\
\alpha_3 &= (P_9u_1 + P_{10}u_2 + P_{11}u_3 + P_{12}u_4)/\Delta \\
\alpha_0 &= (P_{13}u_1 + P_{14}u_2 + P_{15}u_3 + P_{16}u_4)/\Delta
\end{aligned} \tag{4.3}
$$

where Δ is the determinant of the coefficients. Back-substitution of Equations (4.3) into Equation (4.1) gives

$$
\begin{aligned}
u &= (P_1a + P_5b + P_9ab + P_{13})(1/\Delta)u_1 \\
&+ (P_2a + P_6b + P_{10}ab + P_{14})(1/\Delta)u_2 \\
&+ (P_3a + P_7b + P_{11}ab + P_{15})(1/\Delta)u_3 \\
&+ (P_4a + P_8b + P_{12}ab + P_{16})(1/\Delta)u_4
\end{aligned} \tag{4.4}
$$

which in standard form is

$$
u = N_1u_1 + n_2u_2 + N_3u_3 + N_4u_4 \tag{4.5}
$$

Numerical evaluation of the interpolating N-functions using the corner coordinate values of the generic quadrilateral shown in Figure 4.1 leads to:

$$
\begin{aligned}
N_1 &= (1/4)(1 - a - b + ab) = (1/4)(1 - a)(1 - b) \\
N_2 &= (1/4)(1 + a - b - ab) = (1/4)(1 + a)(1 - b) \\
N_3 &= (1/4)(1 + a + b + ab) = (1/4)(1 + a)(1 + b) \\
N_4 &= (1/4)(1 - a + b - ab) = (1/4)(1 - a)(1 + b)
\end{aligned} \tag{4.6}
$$

These N-functions have the property of being equal to 1 at an associated node, but are zero at other nodes, as can be verified by direct substitution of the corner coordinates $(-1, -1)$, $(+1, -1)$, $(+1, +1)$, and $(-1, +1)$ for nodes 1, 2, 3, and 4, respectively. They also have the property that the sum of the four interpolation functions is 1. The same is true of the interpolation functions for the triangle. This property is essential to the representation of a constant displacement. Suppose that, indeed, the node displacements are the same, $u_1 = u_2 = u_3 = u_4 = U$, then according to (4.5), $u = (N_1 + N_2 + N_3 + N_4) U$. Thus, the sum of the interpolation functions must certainly be 1. This observation also shows that derivatives will be zero in the event of a constant displacement field.

4.2 THE ISOPARAMETRIC 4-NODE QUADRILATERAL

A square element would be severely limited in applications. A much more useful element is a quadrilateral illustrated in Figure 4.2.

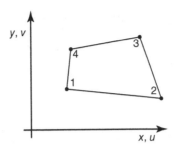

Figure 4.2 A 4-node quadrilateral in the physical (x,y) plane.

Coordinates (x, y) in the physical plane of the quadrilateral in Figure 4.2 may be computed in a manner similar to the computation of displacements in the generic quadrilateral in Figure 4.1, that is, by

$$x = S_1 x_1 + S_2 x_2 + S_3 x_3 + S_4 x_4$$
$$y = S_1 y_1 + S_2 y_2 + S_3 y_3 + S_4 y_4$$
(4.7)

where S_1, S_2, S_3, and S_4 are *shape* functions associated with element geometry in contrast to interpolation functions associated with variables such as solid displacement and fluid pressure.

When the element shape functions are chosen to be the same as the element interpolation functions, the result is called an *isoparametric* element. An apparent advantage to an isoparametric formulation is that the shape functions are immediately known, once the interpolation functions are computed. There is no need for additional computation.

Although the interpolation functions are readily determined in generic (a, b) coordinates, they are actually needed in physical (x, y) coordinates. The connection between the two coordinate systems is a coordinate transformation. Symbolically, the coordinate transformations between (x, y) and (a, b) are

$$x = F(a,b), \quad y = G(a,b)$$
$$a = f(x,y), \quad b = g(x,y)$$
(4.8)

In the isoparametric formulation, the coordinate transformation Equations (4.8) is effected simply by using the shape (interpolation) function for x, y in terms of a, b. Thus,

$$x = N(a,b)_1 x_1 + N(a,b)_2 x_2 + N(a,b)_3 x_3 + N(a,b)_4 x_4$$
$$y = N(a,b)_1 y_1 + N(a,b)_2 y_2 + N(a,b)_3 y_3 + N(a,b)_4 y_4$$
(4.9)

where the dependency on generic coordinates is emphasized. The corner point coordinates x_1, y_1, and so on are constants in Equations (4.9). Intuitively, one expects that the center of the generic quadrilateral (Figure 4.1) should map to the center of the physical quadrilateral (Figure 4.2). In fact, this is the case. The origin of coordinates in the generic plane is at the center of the element where $a=0$, $b=0$. Substitution of these values into Equations (4.9) via Equation (4.5) shows that the corresponding point in the physical plane has coordinates: $x = (0.25) (x_1 + x_2 + x_3 + x_4)$, $y = (0.25) (y_1 + y_2 + y_3 + y_4)$, that are, indeed, the coordinates of the center of the physical quadrilateral. The transformation Equations (4.9) map the generic quadrilateral (Figure 4.1) into the physical quadrilateral (Figure 4.2). This novel coordinate transformation has proven to be very useful in finite element technology.

Chapter 5

Derivatives for a linear displacement quadrilateral

Computation of strains requires expressions for displacement derivatives. The strain–displacement relations are definitions and remain unchanged with the adoption of the isoparametric 4-node quadrilateral. However, the displacement interpolation functions that are needed in x, y coordinates are only available in a, b coordinates as in Equations (4.6). Displacement derivatives required for the strains in x, y coordinates must then be obtained through coordinate transformation and the chain rule for differentiation. Programming for the B-matrix is then straightforward.

5.1 CHAIN RULE APPLICATION

Differentiation of the displacements u, v expressed in standard form gives the normal strains:

$$\varepsilon_{xx} = \partial u/\partial x = (\partial N_1/\partial x)u_1 + (\partial N_2/\partial x)u_2 + (\partial N_3/\partial x)u_3 + (\partial N_4/\partial x)u_4$$
$$\varepsilon_{yy} = \partial v/\partial y = (\partial N_1/\partial y)v_1 + (\partial N_2/\partial y)v_2 + (\partial N_3/\partial y)v_3 + (\partial N_4/\partial y)v_4 \tag{5.1}$$

The shear strain γ_{xy} is obtained in a similar manner, but no additional derivatives of the interpolation functions (N-functions) are needed. However, the interpolation functions are given explicitly in the a, b system and are only known implicitly in the x, y system. Application of the chain rule for partial differentiation resolves the difficulty.

Partial differentiation of N_1 in the a, b system using the chain rule gives

$$\frac{\partial N_1}{\partial a} = \frac{\partial N_1}{\partial x}\frac{\partial x}{\partial a} + \frac{\partial N_1}{\partial y}\frac{\partial y}{\partial a}$$
$$\frac{\partial N_1}{\partial b} = \frac{\partial N_1}{\partial x}\frac{\partial x}{\partial b} + \frac{\partial N_1}{\partial y}\frac{\partial y}{\partial b} \tag{5.2}$$

The remaining interpolation functions are differentiated in the same manner. In matrix form, the results are

$$\begin{Bmatrix} \dfrac{\partial N_1}{\partial a} \\ \dfrac{\partial N_1}{\partial b} \end{Bmatrix} = \begin{bmatrix} \dfrac{\partial x}{\partial a} & \dfrac{\partial y}{\partial a} \\ \dfrac{\partial x}{\partial b} & \dfrac{\partial y}{\partial b} \end{bmatrix} \begin{Bmatrix} \dfrac{\partial N_1}{\partial x} \\ \dfrac{\partial N_1}{\partial y} \end{Bmatrix} \tag{5.3}$$

DOI: 10.1201/9781003166283-5

Similar relations hold for the remaining three interpolation functions. In this regard, the 2×2 matrix in Equation (5.3) is the *Jacobian* of the transformation and is often expressed as $[J]$.

Inversion of Equation (5.3) leads to the desired derivatives. Thus,

$$
\begin{Bmatrix} \dfrac{\partial N_1}{\partial x} \\[6pt] \dfrac{\partial N_1}{\partial y} \end{Bmatrix} = [J]^{-1} \begin{Bmatrix} \dfrac{\partial N_1}{\partial a} \\[6pt] \dfrac{\partial N_1}{\partial b} \end{Bmatrix}
\tag{5.4}
$$

Again, similar relations hold for derivatives of the remaining interpolation functions.

Differentiation of the shape functions that interpolate coordinates allows for determination of the four partial derivatives in the Jacobian matrix. The result in matrix form is

$$
[J]_{(2\times2)} = \begin{bmatrix} \dfrac{\partial N_1}{\partial a} & \dfrac{\partial N_2}{\partial a} & \dfrac{\partial N_3}{\partial a} & \dfrac{\partial N_4}{\partial a} \\[8pt] \dfrac{\partial N_1}{\partial b} & \dfrac{\partial N_2}{\partial b} & \dfrac{\partial N_3}{\partial b} & \dfrac{\partial N_4}{\partial b} \end{bmatrix} \begin{bmatrix} x_1 & y_1 \\ x_2 & y_2 \\ x_3 & y_3 \\ x_4 & y_4 \end{bmatrix}
\tag{5.5}
$$

In an actual computation, the corner coordinates x_1, y_1, and so forth would be known. Evaluation of the matrix containing the partial derivatives in Equation (5.5) is also straightforward and particularly easy when the center of generic quadrilateral is chosen as the point of interest, a rather natural choice. For example, with reference to the derivatives of N_1, one obtains

$$
\frac{\partial N_1}{\partial a} = (-1/4)\,(1 - b), \quad \frac{\partial N_1}{\partial b} = (-1/4)\,(1 - a)
\tag{5.6}
$$

5.2 STRAIN–DISPLACEMENT MATRIX

One form of the strain–displacement matrix for the isoparametric quadrilateral when organized by components of displacement is

$$
\begin{Bmatrix} \varepsilon_{xx} \\ \varepsilon_{yy} \\ \gamma_{xy} \end{Bmatrix} = \begin{bmatrix} P_1 & P_2 & P_3 & P_4 & 0 & 0 & 0 & 0 \\ 0 & 0 & 0 & 0 & P_5 & P_6 & P_7 & P_8 \\ P_5 & P_6 & P_7 & P_8 & P_1 & P_2 & P_3 & P_4 \end{bmatrix} \begin{Bmatrix} u_1 \\ u_2 \\ u_3 \\ u_4 \\ v_1 \\ v_2 \\ v_3 \\ v_4 \end{Bmatrix}
\tag{5.7}
$$

An alternative organization by nodes is an option, of course. In either case, a compact form is

$$
\{\varepsilon\}_{(3\times1)} = [B]_{(3\times8)}\{\delta\}_{(8\times1)}
\tag{5.8}
$$

```
c
c.................for the quadrilateral element nel
c
c.................initialize generic
            corners data aa/-
            1,+1,+1,-1/
            data bb/-1,-1,+1,+1/
c
c.................initialize physical
            corners do 100 j=1,4
            n=nod(nel,j)
            x(j)=xord(n)
100
y(j)=yord(n
) c
c.................initialize generic
            derivatives do 110 j=1,4
            da(j)=0.
25*aa(j       110
db(j)=0.25*bb(j
) c
            do 120 m=1,2
            do 120 n=1,2
            aj(m,n)=0.0
120      bj(m,n)=0.0
c
c.................form the
            Jacobian do 130
            j=1,4
            aj(1,1)=aj(1,1)+d
            a(j)*x(j
            aj(1,2)=aj(1,2)+da(j)*y(j)
            aj(2,1)=aj(2,1)+db(j)*x(j)

130
aj(2,2)=aj(2,2)+db(j)*y(j
) c
c.................invert [J]
            det=aj(1,1)*aj(2,2)-aj(2,1)*aj(1,2)
            bj(1,1)= aj(2,2)/det
            bj(1,2)= -aj(1,2)/det
            bj(2,1)= -aj(2,1)/det
            bj(2,2)= aj(1,1)/det
c
c.................do
            derivatives
            do 140 j=1,4
            P(j)  =bj(1,1)*da(j)+bj(1,2)*db(j)
140      P(j+4)
=bj(2,1)*da(j)+bj(2,2)*db(j)
c
c.................initialize matrix-B
            do1
            50j=
            1,4B
            (1,j)
            =P(j
            )
            B(1,j+4)=0.0
            B(2,j) =0.0
            B(2,j+4)=P(j+4)
            B(3,j) =P(j+4)
150      B(3,j+4)=P(j)
```

Figure 5.1 Program outline for a 4-node isoparametric quadrilateral *B*-matrix (nodal displacement-to-element strain transformation) used in stress analysis.

which is similar in appearance to previous node displacement-to-element strain transformations involving a B-matrix.

The P-numbers are just the derivatives of the interpolation functions in the xy-plane, so

$$P_1 = \partial N_1/\partial x, \ P_2 = \partial N_2/\partial x, \ P_3 = \partial N_3/\partial x, \ P_4 = \partial N_4/\partial x$$
$$P_5 = \partial N_1/\partial y, \ P_6 = \partial N_2/\partial y, \ P_7 = \partial N_3/\partial y, \ P_8 = \partial N_4/\partial y \tag{5.9}$$

In seepage, pressure or head gradients are of interest and are in the role of strains because they appear as derivatives; pressures or heads are in the role of displacements. Interpolation of node pressures over a linear quadrilateral has the form

$$\begin{Bmatrix} \partial p/\partial x \\ \partial p/\partial y \end{Bmatrix} = \begin{bmatrix} P_1 & P_2 & P_3 & P_4 \\ P_5 & P_6 & P_7 & P_8 \end{bmatrix} \begin{Bmatrix} p_1 \\ p_2 \\ p_3 \\ p_4 \end{Bmatrix} \tag{5.10}$$

In compact form, one has

$$\{h\}_{(2\times1)} = [B_p]_{(2,4)}\{H\}_{(4\times1)} \tag{5.11}$$

in terms of gradients and heads.

5.3 PROGRAMMING COMMENTS

Programming the B-matrix quadrilateral for stress analysis requires somewhat greater effort than the triangle. An outline is shown in Figure 5.1. Communication between other parts of a complete finite element program would be through a suitable common block and to some extent through subroutine arguments. Division by the determinant formed in computing the inverse of the Jacobian calls for a trap to warn of a division by zero. One should not rely on the computer for such notification; overflow and underflow are not handled the same on all computers. Some set the result to zero and proceed without warning. Only slight modification is needed to obtain the B_p-matrix for seepage analysis.

The B-matrix in Figure 5.1 is organized by displacement components. When organized by nodes, the arrangement will be different, of course. In this case, the last do-loop in the figure may be replaced by:

```
do 150 j=1,4
m=2*j-1
n=2*j
B(1,m)=P(j)
B(1,n)=0.0
B(2,m)=P(j+4)
B(3,m)=P(j+4)
150 B(3,n)=P(j)
```

Chapter 6

Element equilibrium and stiffness

The purely geometric features of interpolation and strain–displacement relations in stress analysis that were developed intuitively in discrete form now need to be placed within the context of elasticity theory. In this context, a brief statement of appropriate physical laws and material behavior is needed in addition to the previous kinematic (geometric) results. Application of the principle of virtual work to the approximate forms of stress and strain obtained from interpolated displacements leads to a finite element form of the governing equations of elasticity.

The original governing equations in continuum form are a model of physical reality, that is, an approximation. In finite element form, the model is numerical and, in discrete form, approximates the physical model. Programming leads to a computer model that approximates the numerical model with finite arithmetic. Word length is of concern to computer models. Convergence and accuracy of equation solvers are the focus of numerical models as such. The physical model is an engineering judgment about the behavior of the material that is made at the outset of an analysis. Confidence in the reliability of computer models and therefore the usefulness of output comes from an understanding of model details and from the accumulation of modeling experience that enables one to recognize questionable results when they occur. In this way, computer models may be used as valuable engineering tools for rapid, accurate, and low-cost design analyses. Trial designs may be tested before proceeding to construction, and numerical experiments may be performed without the need to build expensive prototypes.

6.1 EQUATIONS FROM ELASTICITY

The primary objective of an elastic analysis of stress is the determination of the motion of a mass of interest, within the elastic limit, subject to specified loads. A motion is characterized by the displacements as functions of position and time. The equations of motion in terms of stress follow from consideration of mass conservation and balances of linear and angular momentum. In two dimensions, the stress equations of motion are

$$
\begin{aligned}
\frac{\partial \sigma_{xx}}{\partial x} + \frac{\partial \tau_{xy}}{\partial y} + \gamma_x = \rho a_x \\
\frac{\partial \tau_{yx}}{\partial x} + \frac{\partial \sigma_{yy}}{\partial y} + \gamma_y = \rho a_y
\end{aligned}
\tag{6.1}
$$

DOI: 10.1201/9781003166283-6

where σ is normal stress, τ is shear stress, γ is specific weight of material, ρ is mass density, and a is acceleration. Acceleration terms are important in dynamic (wave propagation) analyses, but are negligible in static (equilibrium) analyses.

The order of subscripts on the shear stress is not important, that is, $\tau_{xy} = \tau_{yx}$; stress is symmetric. The first subscript refers to a direction perpendicular to the surface acted upon by the considered stress; the second subscript refers to the direction of action of the considered stress. Tension is considered positive when the two directions act in the same sense. A shear stress is positive under this convention when the direction of action and the normal direction are in the same sense. Compression is made positive if the direction of action and normal direction are opposite in sense. The sign convention associated with the double-subscript notation is shown in Figure 6.1.

An important formula in elasticity theory relates stress at the boundary of a body to the surface tractions at the point of interest; this is the famous Cauchy stress formula:

$$T_x = \sigma_{xx}n_x + \tau_{yx}n_y + \tau_{zx}n_z$$
$$T_y = \tau_{xy}n_x + \sigma_{yy}n_y + \tau_{zy}n_z \tag{6.2}$$
$$T_z = \tau_{xz}n_x + \tau_{yz}n_y + \sigma_{zz}n_z$$

where T is a stress vector, that is, a surface traction (force per unit area), and n is a vector of unit length acting normal to the surface at the point of interest, as shown in three dimensions in Figure 6.2.

Kinematics and physical laws lead to fewer equations than unknowns and are not generally sufficient to describe the motion of a deformable body. However, specific characterization of a material response to load, a material law, supplies the deficit information. Thus, material laws or constitutive equations provide additional equations in the form of stress–strain relations and complete the system of governing equations without introducing additional unknowns.

By far, the most widely used material law in rock and soil mechanics, and in structural mechanics as well, is Hooke's law of linear elasticity. Hooke's law is a linear relationship between stress and strain at a point:

$$\varepsilon_{xx} = \left(\frac{1}{E}\right)\left[\sigma_{xx} - v\sigma_{yy} - v\sigma_{zz}\right], \quad \gamma_{xy} = \left(\frac{1}{G}\right)\tau_{xy}$$
$$\varepsilon_{yy} = \left(\frac{1}{E}\right)\left[\sigma_{yy} - v\sigma_{zz} - v\sigma_{xx}\right], \quad \gamma_{yz} = \left(\frac{1}{G}\right)\tau_{yz} \tag{6.3}$$
$$\varepsilon_{zz} = \left(\frac{1}{E}\right)\left[\sigma_{zz} - v\sigma_{xx} - v\sigma_{yy}\right], \quad \gamma_{zx} = \left(\frac{1}{G}\right)\tau_{zx}$$

where E = Young's modulus, v = Poisson's ratio, G = shear modulus, and $G = E/[2(1+v)]$ as usual for isotropic materials. The inverted form of (6.3) is

$$\sigma_{xx} = \left[\frac{E}{(1+v)(1-2v)}\right]\left[(1-v)\varepsilon_{xx} + v\varepsilon_{yy} + v\varepsilon_{zz}\right], \quad \tau_{xy} = G\gamma_{xy}$$
$$\sigma_{yy} = \left[\frac{E}{(1+v)(1-2v)}\right]\left[(1-v)\varepsilon_{xyy} + v\varepsilon_{zz} + v\varepsilon_{xxz}\right], \quad \tau_{xy}z = G\gamma_{yz} \tag{6.4}$$
$$\sigma_{zz} = \left[\frac{E}{(1+v)(1-2v)}\right]\left[(1-v)\varepsilon_{zz} + v\varepsilon_{xx} + v\varepsilon_{yy}\right], \quad \tau_{zx} = G\gamma_{zx}$$

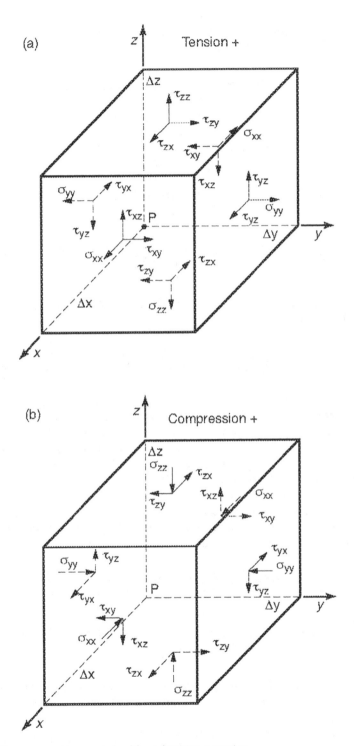

Figure 6.1 Sign convention and double-subscript notation.

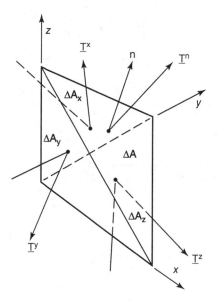

Figure 6.2 Cauchy stress vector.

These equations are also valid in any other locally orthogonal coordinate system, for example, in cylindrical coordinates ($r\theta z$). The only change is replacement of subscripts (xyz) with ($r\theta z$), respectively.

The three-dimensional forms of Hooke's law show that in plane strain analyses (z-direction strains are zero), the z-direction stresses need not be zero, although in isotropic analyses, zero shear strain implies zero shear stress. Similarly, in plane stress analysis, the z-direction stresses are zero, but the corresponding strains need not be zero. Again, in isotropic analyses, zero shear stress implies zero shear strain. If the material is anisotropic, then a zero shear strain does not automatically imply a zero shear stress, nor does a zero shear stress automatically imply a zero shear strain. Thus, even in two-dimensional analyses, consideration of the third dimension may be required.

Recasting (6.3) and (6.4) into matrix notation more suitable for programming and finite element analysis is conveniently done with the aid of a single-subscript notation:

$$\sigma_{xx} = \sigma_1, \ \sigma_{yy} = \sigma_2, \ \sigma_{zz} = \sigma_3, \ \tau_{xy} = \sigma_4, \ \tau_{yz} = \sigma_5, \ \tau_{zx} = \sigma_6$$
$$\varepsilon_{xx} = \varepsilon_1, \ \varepsilon_{yy} = \varepsilon_2, \ \varepsilon_{zz} = \varepsilon_3, \ \gamma_{xy} = \varepsilon_4, \ \gamma_{yz} = \varepsilon_5, \ \gamma_{zx} = \varepsilon_6 \qquad (6.5)$$

Hooke's law in matrix form is then

$$\{\varepsilon\}_{(6\times1)} = [S]_{(6\times6)}\{\sigma\}_{(6\times1)}, \ \ \{\sigma\}_{(6\times1)} = [C]_{(6\times6)}\{\varepsilon\}_{(6\times1)} \qquad (6.6)$$

where the matrices of elastic moduli $[C]$ and compliances $[S]$ are mutual inverses, that is, $[S][C] = [I]$, $[C][S] = [I]$; $[I]$ is the identity matrix with ones on the diagonal and zeroes off the diagonal.

The elastic moduli and compliance matrices are symmetric, so even in the most extreme case of anisotropy, there are at most 21 independent elastic constants. Stratified ground may respond to load parallel to the bedding differently than perpendicular to the bedding and may therefore be considered transversely isotropic and require five elastic constants (Young's moduli parallel and perpendicular to the bedding, Poisson's ratios parallel and perpendicular to the bedding, and an independent cross-bedding shear modulus). Metamorphic rocks may have three distinct material directions and thus may be considered orthotropic and require nine independent elastic moduli (three Young's moduli, three Poisson's ratios, and three shear moduli).

The form of (6.4) needed for two-dimensional finite element programming in plane analyses of isotropic materials is

$$
\begin{Bmatrix} \sigma_{xx} \\ \sigma_{yy} \\ \sigma_{zz} \\ \tau_{xy} \end{Bmatrix} = \begin{bmatrix} 1-v & v & v & 0 \\ v & 1-v & v & 0 \\ v & v & 1-v & 0 \\ 0 & 0 & 0 & (1-2v)/2 \end{bmatrix} \begin{Bmatrix} \varepsilon_{xx} \\ \varepsilon_{yy} \\ \varepsilon_{zz} \\ \gamma_{xy} \end{Bmatrix} \tag{6.7}
$$

And for axial symmetry:

$$
\begin{Bmatrix} \sigma_{rr} \\ \sigma_{\theta\theta} \\ \sigma_{zz} \\ \tau_{rz} \end{Bmatrix} = \begin{bmatrix} 1-v & v & v & 0 \\ v & 1-v & v & 0 \\ v & v & 1-v & 0 \\ 0 & 0 & 0 & (1-2v)/2 \end{bmatrix} \begin{Bmatrix} \varepsilon_{rr} \\ \varepsilon_{\theta\theta} \\ \varepsilon_{zz} \\ \gamma_{rz} \end{Bmatrix} \tag{6.8}
$$

Thus, in both cases which are isotropic, one has a compact form for a plane problem or an axially symmetric problem:

$$
\{\sigma\} = [E]\{\varepsilon\} \tag{6.9}
$$

where $[E]$ is a 4×4 matrix of elastic properties.

6.2 PRINCIPLE OF VIRTUAL WORK

The principle of virtual work can be considered as an engineering application of the well-known divergence theorem. In vector form, the divergence theorem is applicable in any coordinate system and simply states that the integral of the normal component of a vector over the surface of a body is equal to the integral of the divergence of the vector throughout the volume of the body. Thus,

$$
\int_S (U \cdot n)\, dS = \int_V (\nabla \cdot U)\, dV \tag{6.10}
$$

where S is the surface of the body occupying the volume V, U is a vector, n is an outward unit normal vector on S, $(U \cdot n)$ is a dot (inner) product, and $(\nabla \cdot U)$ is the divergence of U in V. In two dimensions,

$$
U \cdot n = U_x n_x + U_y n_y, \quad \nabla \cdot U = \partial U/\partial x + \partial U/\partial y \tag{6.11}
$$

In plane and axially symmetric analyses, only forces acting through the displacements u and v do work, so a two-dimensional formulation of the principle of virtual work is sufficient. In this regard, the product of surface traction T by displacement vector δ is $(T \cdot \delta)$ and is suggestive of the work done by the surface forces (per unit area) applied to the body of interest. The work done by the body force γ (per unit volume) is similarly $(\gamma \cdot \delta)$. Integration over the surface and volume of the body gives

$$W = \int_S (T \cdot \delta)\, dS + \int_V (\gamma \cdot \delta)\, dV \tag{6.12}$$

where W is the total "work" of the applied (external) forces.

The divergence theorem can be used to transform the surface integral in (6.12) to a volume integral after substitution for the surface tractions in terms of the stresses using the Cauchy stress formula (6.2). Thus, in x, y coordinates,

$$
\begin{aligned}
(T \cdot \delta) &= (\sigma_{xx} n_x + \tau_{yx} n_y)u + (\sigma_{xx} n_x + \tau_{yx} n_y)u \\
&= (\sigma_{xx} u + \tau_{yx} v)n_x + (\tau_{xy} u + \sigma_{yy} v)n_y \\
&= (U_x n_x + U_y n_y) \\
(T \cdot \delta) &= (U \cdot n)
\end{aligned}
\tag{6.13}
$$

In view of (6.10) and (6.11),

$$
\begin{aligned}
\int_S T \cdot \delta\, dS &= \int_V \left(\partial U/\partial x + \partial U/\partial y \right) + dV \\
&= \int_V \left[\frac{\partial(\sigma_{xx} u + \tau_{xy} v)}{\partial x} + \frac{\partial(\tau_{yx} u + \sigma_{yy} v)}{\partial y} \right] dV \\
\int_S T \cdot \delta\, dS &= \int_V \left[u\left(\frac{\partial \sigma_{xx}}{\partial x} + \frac{\partial \tau_{yx}}{\partial y} \right) + v\left(\frac{\partial \tau_{xy}}{\partial x} + \frac{\partial \sigma_{yy}}{\partial y} \right) \right] dV \\
&\quad + \int_V \left[\sigma_{xx} \frac{\partial u}{\partial x} + \sigma_{yy} \frac{\partial v}{\partial y} + \tau_{xy} \frac{\partial v}{\partial x} + \tau_{yx} \frac{\partial u}{\partial y} \right] dV
\end{aligned}
\tag{6.14}
$$

Addition of the body force integral in (6.12) to (6.14) leads to

$$
\begin{aligned}
\int_S T \cdot \delta\, dS + \int_V \gamma \cdot \delta\, dV &= \int_V \left[u\left(\frac{\partial \sigma_{xx}}{\partial x} + \frac{\partial \tau_{yx}}{\partial y} + \gamma_x \right) + v\left(\frac{\partial \tau_{xy}}{\partial x} + \frac{\partial \sigma_{yy}}{\partial y} + \gamma_y \right) \right] dV \\
&\quad + \int_V \left[\sigma_{xx} \varepsilon_{xx} + \sigma_{yy} \varepsilon_{yy} + \tau_{xy} \gamma_{xy} \right] dV
\end{aligned}
\tag{6.15}
$$

where the strain–displacement relations have been used in the second integral on the right-hand side. When the system is in equilibrium, the first integral on the right-hand side vanishes because the terms in parentheses are zero. Otherwise, they are the inertia terms as 6.1 shows. Hence,

$$\int_S (T_x u + T_y v)\, dS + \int_V (\gamma_x u + \gamma_y v)\, dV = \int_V [\sigma_{xx} \varepsilon_{xx} + \sigma_{yy} \varepsilon_{yy} + \tau_{xy} \gamma_{xy}]\, dV \tag{6.16}$$

In matrix form, (6.16) is

$$\int_S \begin{Bmatrix} u \\ v \end{Bmatrix}^T \begin{Bmatrix} T_x \\ T_y \end{Bmatrix} dS + \int_V \begin{Bmatrix} u \\ v \end{Bmatrix}^T \begin{Bmatrix} \gamma_x \\ \gamma_y \end{Bmatrix} dV = \int_V \{\varepsilon\}^t \{\sigma\} dV \tag{6.17}$$

where superscript T means transpose of a column matrix to a row matrix.

The same analysis carried out in cylindrical coordinates leads to a similar result and can be obtained by substitution of r and z for x and y in (6.17). The circumferential terms appear when the stress equations of equilibrium, the divergence theorem, and the strain–displacement relations are expressed in cylindrical coordinates.

6.3 ELEMENT EQUILIBRIUM

Element equilibrium is at the core of finite element technology. Although there are several ways of obtaining the element equilibrium equations, application of the principle of virtual work in discrete form is the simplest and most direct way. The discrete form of the principle of virtual work is obtained by simply substituting the interpolated expressions for displacement and strain in (6.17). Thus,

$$\{\delta\}^T \left(\int_S [N]^T \{T\} dS + \int_V [N]^T \{\gamma\} dV - \int_V [B]^T \{\sigma\} dV \right) = 0 \tag{6.18}$$

where the nodal displacements of the considered element $\{\delta\}$ are constants brought outside of the integral signs and the volume integral on the right-hand side of (6.17) is moved to the left-hand side.

If (6.18) is to remain true for arbitrary displacements $\{\delta\}$ from the equilibrium state, then the integral expression within the parentheses in (6.18) must vanish. Hence,

$$\int_S [N]^T \{T\} dS + \int_V [N]^T \{\gamma\} dV = \int_V [B]^T \{\sigma\} dV \tag{6.19}$$

Substitution of Hooke's law and the strain–displacement relations into the right-hand side of (6.19) leads to

$$\int_V [B]^T \{\sigma\} dV = \int_V [B]^T [E] \{\varepsilon\} dV = \left(\int_V [B]^T [E][B] dV \right) \{\delta\} \tag{6.20}$$

where again the nodal forces can be brought outside of the integral sign because they are constants. Equation (6.20) in compact form is:

$$\left(\int_V [B]^T [E][B] dV \right) \{\delta\} = [k] \{\delta\} \tag{6.21}$$

The matrix $[k]$ is an element stiffness matrix, which is a square matrix with dimensions equal to row dimension of the element displacement matrix $\{\delta\}$.

The surface and body force integrals on the left-hand side of (6.21) can be expressed more compactly as a column matrix of forces. Thus,

$$\int_S [N]^T \{T\} dS + \int_V [N]^T \{\gamma\} dV = \{f\} \tag{6.22}$$

Finally, element equilibrium is obtained in the simple form

$$\{f\} = [k]\{\delta\} \tag{6.23}$$

When (6.23) is solved for the element node displacements $\{\delta\}$ under specified forces $\{f\}$, the strains $\{\varepsilon\}$ can be found through $[B]\{\delta\}$; the stresses $\{\sigma\}$ can then be obtained through Hooke's law $[E]\{\varepsilon\}$.

Of course, one element would not be a very satisfactory model of an extended region, so the behavior of an assembly of elements must be determined. However, the processes for computing element strains and stresses will be the same following computation of displacements at all element nodes.

Chapter 7

Global equilibrium and global stiffness

Global equilibrium is required of an assembly of elements that represent a region of interest. Assembly of elements is accomplished essentially by addition of element node forces to obtain the total force at a node in the assembly.

7.1 GLOBAL EQUILIBRIUM

At a typical node i where the element node force is f_i, the total node force F_i is simply

$$F_i = \sum_e f_i \tag{7.1}$$

where the summation is over all elements e that share node i in the assembly. In view of the element equilibrium requirement, one has

$$F_i = \sum_e \sum_{j=1}^{j=n} k_{ij} u_j \tag{7.2}$$

where the inner summation is over the n nodes of the eth element. Elements in two dimensions may be triangles or quadrilaterals ($n = 3$ or 4). After summation is applied to all nodes (elements) and the forces are arranged in a column matrix from the first to last node, the result in compact form is the global equilibrium requirement:

$$\{F\} = [K]\{U\} \tag{7.3}$$

where $\{F\}$, $[K]$, and $\{U\}$ represent a global column matrix of node forces, a global stiffness matrix, and a global column matrix of node displacements.

DOI: 10.1201/9781003166283-7

7.2 GLOBAL ASSEMBLY

The actual assembly process is mainly shifting stiffnesses from local to global positions followed by addition. The local row and column numbers are given global values through an array NOD(*nel*, *j*), where the value of the expression is the global node number of an element *nel* with local node *j*. This array is usually generated automatically by a special program for doing so in advance of an analysis.

A small two-element example illustrates the assembly process. Figure 7.1 shows the two-element array before and after assembly with local and global node numbers.

For element 1 in Figure 7.1,
$$\text{NOD}(1, 1) = 1$$
$$\text{NOD}(1, 2) = 3$$
$$\text{NOD}(1, 3) = 4$$
$$\text{NOD}(1, 4) = 2$$
And for element 2,
$$\text{NOD}(2, 1) = 3$$
$$\text{NOD}(2, 2) = 5$$
$$\text{NOD}(2, 3) = 4$$
$$\text{NOD}(2, 4) = 0$$

There is no 0 numbered node in the global assembly or 4th node of the triangle, so the value of the array NOD(2, 4) is set to 0. As each element is considered in turn, the element forces are added to corresponding global forces. In this example, the global forces can be obtained by inspection. Thus,

$$F_1 = f_1^1$$
$$F_2 = f_4^1$$
$$F_3 = f_2^1 + f_1^2 \tag{7.4}$$
$$F_4 = f_3^1 + f_3^2$$
$$F_5 = f_2^2$$

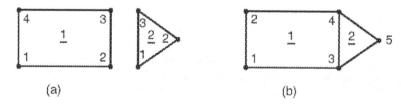

Figure 7.1 Local and global node numbering before and after assembly: (a) before and (b) after.

where the superscript is the element number and the subscript is the element (local) node number. The forces in (7.4) are 2×1 column matrices because they contain x and y components. Subscripts on the left-hand side of (7.4) correspond to global node numbers. The global node number corresponding to local node number 2 of element 1 is 3; the global node number corresponding to local node 1 of element 2 is also 3. The corresponding element forces are added to the global node force at global node 3. These forces obtained are related to element stiffnesses, of course. Hence,

$$
\begin{aligned}
f_1^1 &= k_{11}^1 u_1 + k_{12}^1 u_2 + k_{13}^1 u_3 + k_{14}^1 u_4 \\
f_4^1 &= k_{41}^1 u_1 + k_{42}^1 u_2 + k_{43}^1 u_3 + k_{44}^1 u_4 \\
f_2^1 &= k_{21}^1 u_1 + k_{22}^1 u_2 + k_{23}^1 u_3 + k_{24}^1 u_4 \\
f_1^2 &= k_{11}^2 u_1 + k_{12}^2 u_2 + k_{13}^2 u_3 + 0 \\
f_3^1 &= k_{11}^3 u_1 + k_{12}^3 u_2 + k_{13}^3 u_3 + k_{14}^3 u_4 \\
f_3^2 &= k_{31}^2 u_1 + k_{32}^2 u_2 + k_{33}^2 u_3 + 0 \\
f_2^2 &= k_{21}^2 u_1 + k_{22}^2 u_2 + k_{23}^2 u_3 + 0
\end{aligned}
\tag{7.5}
$$

where the forces f and displacements u are 2×1 column matrices of x and y components and the stiffnesses k are 2×2 matrices.

The row positions of the global forces and therefore of the element stiffness contributions to the global stiffness matrix are given in (7.4). The column position of an element stiffness contribution to the global stiffness matrix is given by the correspondence between local and global displacements and is obtained from the array NOD(*nel*, j). In this example, the global stiffness matrix in terms of element stiffnesses is:

$$
\begin{aligned}
&K_{11} = k_{11}^1 \quad K_{12} = k_{14}^1 \quad K_{13} = k_{12}^1 \quad K_{14} = k_{13}^1 \quad K_{15} = 0 \\
&K_{21} = k_{41}^1 \quad K_{22} = k_{44}^1 \quad K_{23} = k_{42}^1 \quad K_{24} = k_{43}^1 \quad K_{25} = 0 \\
&K_{31} = k_{21}^1 \quad K_{32} = k_{24}^1 \quad K_{33} = k_{22}^1 + k_{11}^2 \quad K_{34} = k_{23}^1 + k_{13}^2 \quad K_{35} = k_{12}^2 \\
&K_{41} = k_{31}^1 \quad K_{42} = k_{34}^1 \quad K_{43} = k_{32}^1 + k_{31}^2 \quad K_{44} = k_{33}^1 = k_{33}^2 \quad K_{45} = k_{32}^2 \\
&K_{51} = 0 \quad\;\; K_{52} = 0 \quad\;\; K_{53} = k_{21}^1 \quad K_{54} = k_{23}^2 \quad K_{55} = k_{22}^2
\end{aligned}
\tag{7.6}
$$

Inspection of (7.6) shows that if the element stiffness matrices are symmetric, then so is the global stiffness matrix. Because a plane (xy) analysis is under consideration, the matrices in (7.6) are 2×2 matrices.

7.3 PROGRAMMING COMMENT

An outline of programming for assembly of the global stiffness matrix is given in Figure 7.2.

The factor of 2 in the definition of row and column positions accounts for the two components of force and displacement at each node. For example, at node 3 in

```
c.....do for all elements
    do 200 j=1,n
    mr=2*nod(nel,j)
    do 200 k=1,n
    nc=2*nod(nel,k)
    SM(mr-1,nc-1)=SM(mr-1,nc-1) +SK(j,k)
    SM(mr-1,nc)  =SM(mr-1,nc)    +SK(j,k+n)
    SM(mr,nc-1)  =SM(mr,nc-1)    +SK(j+n,k)
200 SM(mr,nc)    =SM(mr,nc)      +SK(j+n,k+n)
```

Figure 7.2 An outline of programming for assembly of a global stiffness matrix SM from element stiffness matrices SK.

Figure 7.1, the y force component is at row 5 and similarly for the displacements. In general, if N is the number of nodes in an assembly, then

$$\{F\} = \begin{Bmatrix} F_{1x} \\ F_{1y} \\ F_{2y} \\ F_{2y} \\ \dots \\ F_{Nx} \\ F_{Ny} \end{Bmatrix}_{(2Nx1)} , \quad \{\Delta\} = \begin{Bmatrix} U_{1x} \\ U_{1y} \\ U_{2y} \\ U_{2y} \\ \dots \\ U_{Nx} \\ U_{Ny} \end{Bmatrix}_{(2Nx1)} , \quad [K]_{(2Nx2N0)} \tag{7.7}$$

One should note that the assignment or storage scheme in Figure 7.2 assumes a B-matrix organized by components and *not* by nodes.

When the global stiffness matrix is symmetric, only one half of the off-diagonal stiffnesses need to be saved. A more efficient storage scheme is therefore possible, and that is one that saves only the diagonal terms and, say, the terms above the diagonal stiffnesses. Storage of a master stiffness matrix is then by nodes. For example,

$$\begin{bmatrix} K_{22} & K_{23} & K_{24} & 0 & 0 \\ K_{33} & K_{34} & K_{35} & 0 & 0 \\ K_{44} & K_{45} & 0 & 0 & 0 \\ K_{55} & 0 & 0 & 0 & 0 \end{bmatrix} \tag{7.8}$$

As before, the stiffnesses in (7.8) are 2×2, so there are terms below the main diagonal.

A storage scheme that places the x force component of node 1 in the first row and the y force component in the second row, and so on, is somewhat more compact and is outlined in Figure 7.3.

There is clearly no need to save the column of zeroes on the right in (7.8). In fact, element nodes that are not connected make no contribution to the global or master stiffness matrix. Adjacent nodes are connected; these are nodes of elements that share a node. In this example, node 1 and node 5 are not connected, nor are nodes 2 and 5 in Figure 7.1. In a large assembly of elements, there will be many unconnected nodes

```
              do 200 j=1,n
              jp=2*j
              LP(jp)=2*nod(nel,j)
     200      LP(jp-1)=Lp(jp)-1
     c
              do 210 j=1,nn
              mr=LP(j)
              do 210 k=1,nn
              nc=LP(k)
              if((nc.lt.mr) go to 210
              kc=nc-mr+1
              SM(mr,kc)=SM(mr,kc)+SK(j,k)
     210      continue
```

Figure 7.3 An outline for assembly of a symmetric global stiffness matrix SM from element stiffness matrices SK.

and many zeroes in the global stiffness matrix. In this regard, the greatest difference between adjacent or connected node numbers is indicative of the number of columns beyond which zero entries occur. The number of columns required for storage of the global stiffness matrix is thus reduced to this *bandwidth* (actually, the half-bandwidth in case of symmetry).

The node bandwidth of the master stiffness matrix can be found by the statements in Figure 7.4, where mbd is the bandwidth and nelem is the number of elements in the assembly. The actual bandwidth will be twice the node bandwidth mbd because of the two components of force and displacement in this two-dimensional view. In three dimensions, the bandwidth would be three times the node bandwidth of the global stiffness matrix.

The banded storage scheme may result in many zeroes within the bandwidth. For example, an element with global node numbers 101, 102, 103, and 200, in an assembly of 4-node quadrilaterals, has a node bandwidth of 100 $(200 - 100 + 1)$, but only requires storage of 9 node stiffnesses. The reason is that any one node is only connected to 8 other nodes. Attention should therefore be paid to node numbering to reduce bandwidth and related storage requirements. Figure 7.5 illustrates the advantage of doing so.

An even more efficient storage scheme is the one that stores only non-zero stiffnesses and eliminates zeroes within the master stiffness matrix bandwidth. Figure 7.6 outlines programming for such a scheme, which has the advantage of minimizing bandwidth while allowing for arbitrary global numbering of nodes.

```
              mbd=0
              do 150 i=1,nelem
              m=nod(I,1)
              do 150 j=2,n
              n=nod(I,j)
              if(n.eq.0) go to 150
                nd=iabs(m-n)+1
              if (nd.gt.mbd) then
                mbd=nd
              endif
     150      continue
```

Figure 7.4 An outline of programming for node semi-bandwidth determination.

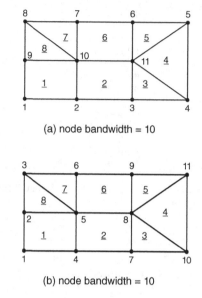

(a) node bandwidth = 10

(b) node bandwidth = 10

Figure 7.5 Alternative node numbering schemes and relationship to bandwidth.

```
c..............initialize arrays
        do 205 m=1,nnode
        nap(m)=0
        do 200 jc=1,ndj
200     np(m,jc)=0
205     np(m,1)=m
c....................do for all elements
        do 250 j=1,n
        m=nod(nel,j)
        mr=2*m
        do 250 k=1,n
        n=nod(nel,k)
        jc=0
210     jc=jc+1
        If(np(m,jc).eq.n) to to 220
        If(np(m,jc).ne.0) to to 210
220     np(m,jc)=n
        kc=2*jc
        SM(mr-1,kc-1)=Sm(mr-1,kc-1)+SK(j,k)
        SM(mr-1,kc)  =Sm(mr-1,kc)  +SK(j,k+n)
        SM(mr,kc-1)  =Sm(mr,kc-1)  +SK(j,k)
        SM(mr,kc)    =Sm(mr,kc)    +SK(j+n,k+n)
c
c..............count connected nodes
        do 300 m=1,nnode
        do 300 jc=1,ndj
        if(np(m,jc,gtl0) then
        nap(m)=nap(m)+1
        endif
300     continue
```

Figure 7.6 Programming for storing only non-zero master stiffness coefficients while generating a pointer matrix np(nnode, ndj).

Chapter 8

Static condensation and a 4CST element

A quadrilateral can obviously be formed from four triangles. Figure 8.1 shows a quadrilateral disassembled into four constant-strain triangles and then reassembled as a quadrilateral with a center node. The center node is subsequently removed to form a quadrilateral known as the 4CST finite element. An advantage of the 4CST element is explicit formulation; numerical integration is not required. Numerical performance of the 4CST element is far superior to the sub-assembly of four constant-strain triangles, a consequence of eliminating the center node. The center node is also virtual node generated during run time; additional storage is not required. Element input is as a quadrilateral. Elimination of the center node during the computation of the 4CST element leads to the general issue of equation solving.

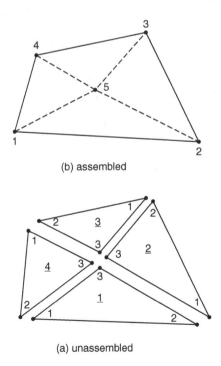

(b) assembled

(a) unassembled

Figure 8.1 A 4CST quadrilateral formed from four constant-strain triangles.

DOI: 10.1201/9781003166283-8

8.1 STATIC CONDENSATION

The computation is simply a process of eliminating unknowns by substitution. Consider the 3×3 system

$$
\begin{aligned}
a_{11}x_1 + a_{12}x_2 + a_{13}x_3 &= b_1 \\
a_{21}x_1 + a_{22}x_2 + a_{23}x_3 &= b_2 \\
a_{31}x_1 + a_{32}x_2 + a_{33}x_3 &= b_3
\end{aligned}
\tag{8.1}
$$

The third equation solved for the third unknown gives

$$
x_3 = \left(\frac{b_3}{a_{33}}\right) - \left(\frac{a_{31}}{a_{33}}\right)x_1 - \left(\frac{a_{32}}{a_{33}}\right)x_2
\tag{8.2}
$$

After substitution into the first two equations, the result is a reduced system of equations. Thus,

$$
\begin{aligned}
\left[a_{11} - a_{13}\left(\frac{a_{31}}{a_{33}}\right)\right]x_1 + \left[a_{12} - a_{13}\left(\frac{a_{32}}{a_{33}}\right)\right]x_2 &= \left[b_1 - a_{13}\left(\frac{b_3}{a_{33}}\right)\right] \\
\left[a_{21} - a_{23}\left(\frac{a_{31}}{a_{33}}\right)\right]x_1 + \left[a_{22} - a_{13}\left(\frac{a_{32}}{a_{33}}\right)\right]x_2 &= \left[b_2 - a_{23}\left(\frac{b_3}{a_{33}}\right)\right]
\end{aligned}
\tag{8.3}
$$

Repetition of the process would reduce the system once again and, in this simple case, would result in the solution of the first unknown. Substitution of this value into the second system of equations would give the second unknown. Substitution of these two unknowns into the third equation of the original system completes the solution process of backward elimination and forward substitution. Usually, the elimination or reduction process is carried forward beginning with the first unknown, followed by a backward substitution process. However, the objective here is to simply "reduce out" the unknowns associated with node 5 in the 4CST sub-assembly. There are two unknown displacements associated with each node, so the 5-node sub-assembly stiffness matrix is (10×10). The reduction process is therefore carried out twice to obtain the (8×8) stiffness matrix of the 4CST quadrilateral.

Forces are not considered, although they could be with slight modification. The center node 5 is contained entirely within the element and would ordinarily be force-free. In this case, the forces at the corner nodes of the quadrilateral are unaffected and no provisions for storage of the center node forces and displacements are necessary.

8.2 PROGRAMMING COMMENT

Programming requires computing the stiffness matrix for each triangle and sub-assembly into a 5-node array. The center node is then eliminated before passing the new 4-node quadrilateral stiffness matrix to a global assembly subroutine. Shifting of triangle stiffnesses to the proper position in the 5-node array is a local operation that is the same for every 4CST element. Triangle stiffness is (6×6); the 5-node array is (10×10); and the 4CST element stiffness is (8×8). The global node numbers of a 4CST quadrilateral are defined by input using the array NOD(nel, j), $j = 1, 2, 3, 4$, that is, by the four corner nodes only.

Programming for the 5-node sub-assembly is outlined in Figure 8.2. The array nrc(6, 4) is a "node row–column" array that shifts triangle stiffnesses into the proper 5-node sub-assembly locations. Reduction of the 5-node sub-assembly to the final form

```
c...................initialize row-column array
data nrc/1,2,3,4,9,10,3,4,5,6,9,10, 5,6,7,8,9,10,7,8,1,2,9,10/
c...................fix quadrilateral center
       xc=0.0
       yc=0.0
       do 110 j=1,n
       k=NOD(nel,j)
       x(j)=xord(k)
       y(j)=yord(k)
       xc=xc+x(j)/n
110    yc=yc+y(j)/n
c...................do 4 triangles
c
       do 290 nt=1,4
c
       do 210 mt=1,2
       jp=nt-1+mt
       if(jp.eq.5) jp=1
       k=NOD(nel,jp)
       x(mt)=xord(k)
210    y(mt)=yord(k)

c...................do determinant and coefficients
       P(7) = x(2)*yc-xc*y(2)
       P(8) = xc*y(1)-x(1)*yc
       P(9) = x(1)*y(2)-x(2)*y(1)
       det  = P(7)+P(8)+P(9)
       P(1) = (y(2)-yc)/det
       P(2) = (yc-y(1))/det
       P(3) = (y(1)-y(2))/det
       P(4) = (xc-x(2))/det
       P(5) = (x(1)-xc)/det
       P(6) = (x(2)-x(1))/det
       P(10)=( P(7)/xc+P(1)+P(4)*(yc/xc) )/det
       P(11)=( P(8)/xc+P(2)+P(5)*(yc/xc) )/det
       P(12)=( P(9)/xc+P(3)+P(6)*(yc/xc) )/det
c
c...................initialize [B] by nodes
       do 220 j=1,3
       n=2*j
       m=2*j-1
       B(1,m) =P(j)
       B(1,n) =0.0
       B(2,m) =0.0
       B(2,n) =P(j+3)
       B(3,m) =0.0
       B(3,n) =0.0
       B(4,m) =P(j+3)
220    B(4,n) =P(j)
c
c...................form [E][B]
       do 230 j=1,4
       do 230 k=1,6
       temp(j,k)=0.0
       do 230 m=1,4
230    temp(j,k)=temp(j,k)+E(j,m)*B(m,k)
c
c...................form ∫[B]ᵀ[E][B]dV = [SK]
c
       V=0.5*det
       do 245 j=1,6
       do 245 k=1,6
       s(j,k)=0.0
       do 240 m=1,4
240    s(j,k)=s(j,k)+B(m,j)*temp(m,k)
245    s(j,k)=s(j,k)*V
c
c...................sub-assembly process
c
       do 250 j=1,6
       m=nrc(j,nt)
       do 250 k=1,6
       n=nrc(k,nt)
250    SK(m,n)=SK(m,n)+s(j,k)
290    continue
```

Figure 8.2 Outline for the 4CST element through the 5-node sub-assembly.

of the 4CST quadrilateral stiffness is outlined in Figure 8.3. The reduction process that removes the center node is essentially a step in the equation-solving process that uses an elimination scheme at the element level.

The coefficients $P(10)$, $P(11)$, and $P(12)$ are computed for possible use in axial symmetry, although the B-matrix is initialized for plane analysis.

```
c.................next reduce to quadrilateral
c
        do 305 jk=1,2
        n=10-jk
        m= n+1
        do 305 j=1,n
        SE=SK(m,j)/SK(m,m)
        do 300 k=j,n
        SK(k,j)=SK(k,j)-SE*SK(k,m)
300     SK(j,k)=SK(k,j)
305     continue
```

Figure 8.3 Program outline to eliminate node 5 from the 5-node sub-assembly, leading to the 4CST quadrilateral element stiffness [SK] for stress analysis.

Chapter 9

Equation solving

There are two important processes for solving equations: elimination and iteration. Several varieties of each approach are described in detail in texts on numerical analysis. In finite element applications, elimination seems to be the most popular approach. However, iteration plays an essential role in nonlinear analyses, for example, in elastic–plastic analysis of stress or in analysis of seepage with pressure-dependent hydraulic conductivity. Even when the basic solution scheme is elimination, some form of iteration is usually required. In very large systems of equations (millions of unknowns), iteration is almost certain to be required.

9.1 GAUSS ELIMINATION

Consider the symmetric system of equations:

$$
\begin{bmatrix}
a_{11} & a_{12} & a_{13} & 0 & 0 & 0 & 0 \\
a_{21} & a_{22} & a_{23} & a_{24} & 0 & 0 & 0 \\
a_{31} & a_{32} & a_{33} & a_{34} & a_{35} & 0 & 0 \\
0 & a_{42} & a_{43} & a_{44} & a_{45} & 0 & a_{47} \\
0 & 0 & a_{53} & a_{54} & a_{55} & a_{56} & a_{57} \\
0 & 0 & 0 & 0 & a_{65} & a_{66} & a_{67} \\
0 & 0 & 0 & a_{74} & a_{75} & a_{76} & a_{77}
\end{bmatrix}
\begin{Bmatrix}
x_1 \\ x_2 \\ x_3 \\ x_4 \\ x_5 \\ x_6 \\ x_7
\end{Bmatrix}
=
\begin{Bmatrix}
b_1 \\ b_2 \\ b_3 \\ b_4 \\ b_5 \\ b_6 \\ b_7
\end{Bmatrix}
\tag{9.1}
$$

which has a bandwidth of four. To eliminate the first unknown x_1 from the second row, the first row is divided by the leading coefficient a_{11} and then multiplied by the leading coefficient of the second row a_{21} and finally subtracted from the second row with the result

$$
0 + \left[a_{22} - a_{21}\left(\frac{a_{12}}{a_{11}}\right) \right]x_2 + \left[a_{23} - a_{21}\left(\frac{a_{13}}{a_{11}}\right) \right]x_3 + \left[a_{24} - a_{21}\left(\frac{a_{14}}{a_{11}}\right) \right]x_4
$$
$$
+ \left[a_{25} - a_{21}\left(\frac{a_{15}}{a_{11}}\right) \right]x_5 + \left[a_{26} - a_{21}\left(\frac{a_{16}}{a_{11}}\right) \right]x_6 + \left[a_{27} - a_{21}\left(\frac{a_{17}}{a_{11}}\right) \right]x_7 \tag{9.2}
$$
$$
= b_2 - a_{21}\left(\frac{b_1}{a_{11}}\right)
$$

DOI: 10.1201/9781003166283-9

When account is taken of the zeroes, the result is

$$
0 + \left[a_{22} - a_{21}\left(\frac{a_{12}}{a_{11}}\right) \right] x_2 + \left[a_{23} - a_{21}\left(\frac{a_{13}}{a_{11}}\right) \right] x_3 + \left[a_{24} - a_{21}\left(\frac{a_{14}}{a_{11}}\right) \right] x_4
$$
$$
+ 0 + 0 + 0 = b_2 - a_{21}\left(\frac{b_1}{a_{11}}\right)
$$
(9.3)

The same procedure eliminates x_1 from the third row with the result

$$
0 + \left[a_{32} - a_{31}\left(\frac{a_{12}}{a_{11}}\right) \right] x_2 + \left[a_{33} - a_{31}\left(\frac{a_{13}}{a_{11}}\right) \right] x_3 + \left[a_{34} \right] x_4
$$
$$
+ \left[a_{35} \right] x_5 + 0 + 0 = b_3 - a_{31}\left(\frac{b_1}{a_{11}}\right)
$$
(9.4)

The new coefficients after eliminating x_1 from the seventh row are

$$
a_{ij}^{(1)} = a_{ij} - a_{i1}\left(\frac{a_{1j}}{a_{11}}\right), \qquad (i, j = 2, 3, \ldots, 7)
$$
(9.5)

where the superscript indicates elimination of the first unknown. In this example, the system has seven unknowns, but clearly, (9.5) is valid for an arbitrary number of unknowns.

The entire process is now restarted beginning with the second row and proceeding down the second column to eliminate x_2. After elimination of x_2 from the seventh row, the resulting coefficients are given by

$$
a_{ij}^{(2)} = a_{ij}^{(1)} - a_{i2}^{(1)}\left(\frac{a_{2j}^{(1)}}{a_{22}^{(1)}}\right), \qquad (i, j = 3, 4, \ldots, 7)
$$
(9.6)

In general, after the mth reduction in a system of n unknowns

$$
a_{ij}^{(m)} = a_{ij}^{(m-1)} - a_{i2}^{(m-1)}\left(\frac{a_{2j}^{(m-1)}}{a_{22}^{(m-1)}}\right), \qquad (i, j = m+1, \ldots, n)
$$
(9.7)

Because of the assumed symmetry of the coefficient matrix and the fact that symmetry prevails after each reduction, the equality

$$
a_{im}^{(m-1)} = a_{mi}^{(m-1)}
$$
(9.8)

holds. This fact has been used in (9.7). As a consequence, coefficients below the main diagonal are not required and do not need to be stored. The banded symmetric storage scheme may be used.

The right-hand-side entries (forces) after the mth reduction are

$$b_i^{(m)} = b_i^{m-1} - a_{mi}^{(m-1)} \left(\frac{b_m^{(m-1)}}{a_{mm}^{(m-1)}} \right) \tag{9.9}$$

where symmetry is used.

After elimination of the $n-1$ unknowns, only the nth unknown remains in the last equation. In the example, the result is

$$\begin{bmatrix} a_{11} & a_{12} & a_{13} & 0 & 0 & 0 & 0 \\ 0 & a_{22} & a_{23} & a_{24} & 0 & 0 & 0 \\ 0 & 0 & a_{33} & a_{34} & a_{35} & 0 & 0 \\ 0 & 0 & 0 & a_{44} & a_{45} & 0 & a_{47} \\ 0 & 0 & 0 & 0 & a_{55} & a_{56} & a_{57} \\ 0 & 0 & 0 & 0 & 0 & a_{66} & a_{67} \\ 0 & 0 & 0 & 0 & 0 & 0 & a_{77} \end{bmatrix} \begin{Bmatrix} x_1 \\ x_2 \\ x_3 \\ x_4 \\ x_5 \\ x_6 \\ x_7 \end{Bmatrix} = \begin{Bmatrix} b_1 \\ b_2 \\ b_3 \\ b_4 \\ b_5 \\ b_6 \\ b_7 \end{Bmatrix} \tag{9.10}$$

where the coefficients and forces are the values generated during the elimination process. The last equation is easily solved for the last unknown x_7. Substitution of this value into the next to the last equation allows for solution of the next to the last unknown x_6. Continuation of this backward substitution process proceeds until the first unknown is found.

9.2 ELIMINATION BOUNDARY CONDITIONS

There are two types of boundary conditions to consider in an analysis of stresses, forces, and displacements. In seepage analysis, either fluid pressure or velocity may be specified on a boundary of a region of interest. Forces enter rather naturally into the global system of equilibrium equations as a column matrix $\{F\}$ in the global equilibrium equation $\{F\} = [K]\{U\}$. Displacement boundary conditions require special consideration. In this regard, there should be sufficient displacement boundary conditions to prevent rigid body motion (translation; rotation) of the region of interest. Otherwise, the system will be singular.

Consider the 7×7 example system of equations, and suppose x_2 is specified. As a "known", it may be moved to the right-hand side of the system as shown in (9.11), where the corresponding row and column entries in the global stiffness matrix are replaced by zeroes except at the diagonal position where the entry is 1. Solution of (9.11) will produce all the unknowns and give the proper value of x_2. The same process may be applied to other unknown displacements. Each time, the global stiffness and force matrices are modified in a similar manner. If all the displacements are specified, then of course the solution is at hand and ready for calculating strains and stresses.

In seepage analysis, pressures correspond to displacements; pressure gradients correspond to strains; and velocities correspond to stresses.

$$
\begin{bmatrix}
a_{11} & 0 & a_{13} & 0 & 0 & 0 & 0 \\
a_{21} & 1 & a_{23} & a_{24} & 0 & 0 & 0 \\
a_{31} & 0 & a_{33} & a_{34} & a_{35} & 0 & 0 \\
0 & 0 & a_{43} & a_{44} & a_{45} & 0 & a_{47} \\
0 & 0 & a_{53} & a_{54} & a_{55} & a_{56} & a_{57} \\
0 & 0 & 0 & 0 & a_{65} & a_{66} & a_{67} \\
0 & 0 & 0 & a_{74} & a_{75} & a_{76} & a_{77}
\end{bmatrix}
\begin{bmatrix}
x_1 \\ x_2 \\ x_3 \\ x_4 \\ x_5 \\ x_6 \\ x_7
\end{bmatrix}
=
\begin{Bmatrix}
b_1 - a_{12}x_2 \\
x_2 \\
b_3 - a_{32}x_2 \\
b_4 - a_{42}x_2 \\
b_5 - a_{52}x_2 \\
b_6 - a_{62}x_2 \\
b_7 - a_{72}x_2
\end{Bmatrix}
\tag{9.11}
$$

9.3 GAUSS–SEIDEL ITERATION

Consider a typical equation in a global system in the form

$$
F_i = \sum_{j=1}^{j=i-1} K_{ij}U_j + K_{ii}U_i + \sum_{i=j+1}^{j=m} K_{ij}U_j
\tag{9.12}
$$

where there are m terms in this ith row. Equation (9.12) is readily solved for the un-known U_j. First, the unknown term is moved to the left-hand side and then added and subtracted from the right-hand side:

$$
K_{ii}U_i = K_{ii}U_i + \left[F_i - \sum_{j=1}^{j=i-1} K_{ij}U_j - \sum_{i=j}^{j=m} K_{ij}U_j \right]
\tag{9.13}
$$

An approximate solution for U_i is then

$$
U_i^{(n+1)} = U_i^{(n)} + \left[F_i - \sum_{j=1}^{j=n-1} K_{ij}U_j^{(n+1)} - \sum_{j=n+1}^{j=m} K_{ij}U_j^{(n)} \right] / K_{ii}
\tag{9.14}
$$

where the superscript indicates the iteration number. All the displacements on the right-hand side of (9.14) are the latest estimates from the $n+1$ iteration up to ith dis-placement and from the nth iteration for remaining displacements. Equation (9.14) is more compactly written as

$$
U_i^{(n+1)} = U_i^{(n)} + \alpha R_i / K_{ii}
\tag{9.15}
$$

where R_i is a "residual", a computed difference between the left- and right-hand sides of the equilibrium equation, and α is an *over-relaxation factor* that accelerates conver-gence of the iteration scheme. This particular line iteration scheme is Gauss–Seidel iteration with systematic over-relaxation, in short SOR.

The over-relaxation factor has an optimum value between 1 and 2, but is generally not known. In the absence of an optimum value for the problem at hand, a value of 1.8

is suggested. In this regard, a value less than optimum will simply slow convergence. If convergence is very slow and iterative improvement occurs in digits beyond the word length of the computer, then convergence ceases numerically. A highly overestimated convergence factor may cause oscillatory convergence, where the residuals alternatively increase and decrease. If the "period" of oscillation is long, then the solution may appear to diverge.

Because Gauss–Seidel convergence is not necessarily monotonic, comparisons of successive residuals may not be a useful criterion for exiting from the equation solver. Still, some "error" tolerance needs to be specified, and of course, there should be a limit to the number of iterations allowed for the problem at hand. Error is the difference between the numerical estimate and the true solution. Although the true solution is unknown, residuals behave numerically as the error. In fact, a proper measure of the residuals is a "norm". One easily computed norm is simply the sum of the absolute values of the computed residuals. Monitoring the norm during iteration then allows for exiting the equation solver when the norm is sufficiently small, a matter of choice. Usually, several orders of magnitude reduction in the norm is indicative of a reasonably well-converged solution.

An alternative to solving (9.12) is

$$U_i = \left(1/K_{ii}\right)\left(F_i - \sum_{\substack{j=i \\ j \neq i}}^{j=m} K_{ij}U_j\right) \tag{9.16}$$

The iteration is then

$$U_i^{(n+1)} = \left(1/K_{ii}\right)\left(F_i - \sum_{\substack{j=i \\ j \neq i}}^{j=m} K_{ij}U_j^{(n)}\right) \tag{9.17}$$

This iteration is the well-known *Jacobi* iteration. A disadvantage of Jacobi iteration is the requirement for storing two vectors $U_i^{(n+1)}$ and $U_i^{(n)}$, where i ranges over the unknowns.

Yet another iterative procedure is the *conjugate gradient* method. There are several versions of this equation solver, which has the interesting feature of guaranteed convergence to the exact solution after a finite number of iterations. However, the number of iterations for an exact solution may be enormous. Moreover, satisfactory convergence is usually achieved with far less iterations. Details of this popular method are beyond the scope of these Notes.

9.4 ITERATION BOUNDARY CONDITIONS

Boundary conditions are easily handled using SOR. As before, forces need no special treatment. Specified displacement may be handled by simply skipping the corresponding line in the iteration. Alternatively, the row coefficients and corresponding force may be modified by setting the off-diagonal stiffness to zero and the diagonal stiffness to 1 with the force replaced by the specified displacement. Column coefficients are not modified as in an elimination solver. This alternative reduces the row residual to zero and the row iteration to $U_i^{(n+1)} = U_i^n$, thus leaving the specified displacement U_i unchanged.

9.5 PROGRAMMING COMMENTS FOR ELIMINATION

An outline of programming for forward reduction of the global stiffness and force matrices followed by backward substitution for the unknown displacements is shown in Figure 9.1.

The programming in Figure 9.1 follows the explanation of Gauss elimination rather closely. Although symmetry of the master stiffness matrix is assumed, there are no provisions for avoiding zero multiplications outside the bandwidth of the system. There is also a tacit assumption that the diagonal stiffnesses are not zero, so zero division will not occur. Unnecessary zero multiplications can be avoided by simply limiting the range of the do 210, do 229, do 240, and do 260 loops to *mbd*, the bandwidth of the system. In this regard, the advantage of banded storage can be obtained with only slight modifications of the programming in Figure 9.1.

Figure 9.2 shows a banded equation-solving scheme that follows from these modifications to Figure 9.1. The range of the inner do-loops is the minimum of (*mbd*, *kb*) as determined by the library routine MINO(). This limit prevents subscripts from becoming larger than allowed. For example, during the first pass of the backward substitution process for the next to the last unknown, if the do 260 loop ranged to the bandwidth, then the subscript of the last force term at the end of the do-loop range would be

```
      c.................forward reduction of [SM]
      c
                neq=2*nnode
                do 230 m=1,neq-1
                do 220 i=m+1,neq
                C=SM(m,i)/SM(m,m)
                do 210 j=i,neq
      210       SM(i,j)=SM(i,j)-C*SM(m,j)
      220       SM(m,i)=C
      230       continue
      c.................forward reduction of {F}
      c
                 do 250 m=1,neq-1
                 do 240 i=m+1,neq
      240       F(i)=F(i)-SM(m,i)*F(m)
      250       F(m)=F(m)/SM(m,m)
      c
                 F(neq)=F(neq)/SM(neq,neq)
      c
      c.................back substitution for {U}={F}
      c
                 do 260 i=neq-1,1,-1
                 do 260 j=i+1,neq
      260       F(i)=F(i)-SM(i,j)*F(j)
```

Figure 9.1 Outline of Gauss elimination equation solving (forward elimination and backward substitution).

```
                     ..................forward reduction of [SM]
                          neq=2*nnode
                          do 230 m=1,neq-1
                            kb=neq-m+1
                            mb=MIN0(mbd,kb)
                            do 220 i=2,mb
                            C=SM(m,i)/SM(m,1)
                              k=m+i-1
                              n=0
                              do 210 j=i,mb n=n+1
          210               SM(k,n)=SM(k,n)-C*SM(m,j)
          220             SM(m,i)=C
          230           continue
          c..................forward reduction of {F}
          c
                          do 250 m=1,neq-1
                            kb=neq-m+1
                            mb=MIN0(mbd,kb)
                            do 240 j=2,mb
                              k=m+j-1
          240               F(k)=F(k)-SM(m,j)*F(m)
          250             F(m)=F(m)/SM(m,1)
          c
                          F(neq)=F(neq)/SM(neq,1)
          c
          c..................back substitution for {U}={F}
          c
                          do 260 m=neq-1,1,-1
                            kb=neq-m+1
                            mb=MIN0(mbd,kb)
                            do 260 j=2,mb
                              k=m+j-1
          260               F(m)=F(m)-SM(m,j)*F(k)
```

Figure 9.2 Outline of Gauss elimination equation solving for banded symmetric stiffness matrix storage.

(new+nmbd-1), when it should actually be (new-1). In any case, the unknown displacements appear in place of the original forces, that is, in $\{F\}$ after back-substitution.

Programming that modifies the global stiffness and force matrices according to the example is outlined in Figure 9.3. The last statement in Figure 9.3 substitutes the known displacement for the force at the given position. During back-substitution, as the displacements are computed, the same replacement occurs and the now-known

```
          c..................for specified displacement U at N

          c
                  do 220 m=2,mbd k=N-m+1
                  if(k.le.0) go to 210 F(k)=F(k)-SM(k,m)*U SM(k,m)=0.0
          210     k=N+m-1
                  if(k.gt.neq) go to 220 F(k)=F(k)-SM(N,m)*U SM(N,m)=0.0
          220     continue c
                  SM(N,1)=1.0
                  F(N)=U
          c..........................
```

Figure 9.3 Outline of Gauss elimination equation solving for banded symmetric stiffness storage for displacement boundary conditions.

displacements replace the forces in {F}. The if-statements in Figure 9.3 prevent subscripts from going out of range.

9.6 PROGRAMMING COMMENTS FOR ITERATION

Programming for Gauss–Seidel iteration is outlined in Figure 9.4. Provision is made for writing the residuals to the screen after every *inter* (interval) number of iterations, perhaps 20–40 for a small problem and 100–200 for a large problem. The maximum number of iterations allowed is *maxit*, perhaps 200–400 for a small problem or 2,000 and more for a large problem. Whether convergence is achieved at the maximum number of iterations allowed is a matter of judgment. The effect of increasing the maximum number of iterations and the error allowed can be assessed by rerunning the analysis.

```
        c...................Gauss-Seidel iteration c
                ncycl=inter
                do 250 it=1,maxit
                smr=0.0
                do 230 mr=1,neq
                 jp=nap(mr)
                do 220 j=1,jp
                 k=np(mr,j)
                 res=F(mr)
        220     res=res-SM(mr,j)*U(k)
                 U(k)=U(k)+orf*res/SM(mr,1)
        230     smr=smr+abs(res)
                if(smr.lt.err) then
                   go to 290
                elseif(it.lt.ncycl) then
                   go to 250
                else
                   write(*,*) ' Iterations completed = ',it,
                .  Residuals = ',smr ncycl=ncycl+inter
                   endif
        250     continue
        290     write(*,*) ' Iterations completed = ',it,
                    .        '      Residuals = ',smr
```

Figure 9.4 Outline of Gauss–Seidel line iteration equation solving using over-relaxation with storage of non-zero stiffnesses only.

Chapter 10

Material nonlinearity

Material nonlinearity occurs when properties of the material change significantly under load. For example, in stress analysis, continued loading of a ductile material beyond the elastic limit results in a plastic contribution to the total strain, which then depends on the current state of stress. The slope of the stress–strain curve (E) decreases significantly, and the relationship between stress and strain is no longer linear. In seepage analysis, hydraulic conductivity may depend on fluid pressure. For example, when pressure tends to become negative, surface tension of void water and menisci between water and solid grains comes into play. Desaturation occurs with the result that a much higher pressure gradient is required to move water through the film adjacent to the solid skeleton of the partially saturated rock mass. Resistance to fluid flow is greatly increased, while hydraulic conductivity (k) is greatly reduced under these conditions. In both cases, the onset of material nonlinearity is marked by a limiting value of one of the unknowns (stress; pressure). Beyond the limiting value, the applicable material law assumes a differential form that is also valid in the strictly linear range of response.

There are two common approaches to material nonlinearity in finite element analysis; these are *incremental* and *iterative* approaches. The former is well suited to iterative equation solving (Gauss–Seidel iteration) and is a *tangent stiffness* approach. The latter is generally used with elimination equation solvers (Gauss elimination) and is actually a combination of elimination and iteration (*Newton–Raphson* approach). Both require reassembly of the master stiffness matrix, although a modified Newton–Raphson approach avoids reassembly when the material nonlinearity is not widespread. The advantages of each are linked mainly to the computational efficiencies of the processes for accomplishing this task. The Newton–Raphson approach is more often used to cope with geometric nonlinearity that occurs when displacements and strains become large.

10.1 INCREMENTAL (TANGENT STIFFNESS) APPROACH

In elastic–plastic analysis, the limit to a purely linearly elastic response occurs when the stresses satisfy a yield condition such as the well-known Mohr–Coulomb condition. The differential stress–strain relations in matrix notation are then:

$$\{d\sigma\} = \left[E\big(\{\sigma\}, \{\varepsilon^p\}\big) \right]\{d\sigma\} \tag{10.1}$$

DOI: 10.1201/9781003166283-10

where $\{\varepsilon^p\}$ is the plastic component of strain and $\left[E\big(\{\sigma\},\{\varepsilon^p\}\big) \right]$ is the *tangent stiffness* of the material. Under uniaxial loading, E is the local slope (tangent modulus) of the stress–strain curve. When $\{\varepsilon^p\} = \{0\}$, $[E]$ is simply the 6×6 matrix of elastic moduli and (10.1) is just Hooke's law in differential form. The details for computing $[E(\{\sigma\}, \{p\})]$ are obtained from plasticity theory for geologic media, which is beyond the scope of these Notes.

In seepage analysis, a simplified view is that the fluid "yields" in tension, that is, when the fluid pressure (considered positive) becomes negative. Darcy's law is then:

$$\{dv\} = [k\{p\}]\{dh\} \tag{10.2}$$

When fluid pressure p is compressive, hydraulic conductivity $[k]$ is constant. However, in simplified form (neglect of surface tension), tensile p is not allowed and $[k] = 0$. Hence, $[k\{p\}]$ is an array of step functions.

In the tangent stiffness approach, the differential relations (10.1 and 10.2) are replaced by incremental relations. For example, in stress analysis, (10.1) is

$$\{\Delta\sigma\} = [E(\{\sigma\},\{\varepsilon^p\})]\{\Delta\sigma\} \tag{10.3}$$

The applied loads are then imposed in a number of "small" increments or steps. The actual number of load increments required to achieve a reasonable approximation to the differential nature of the problem is a matter of judgment; ten load steps are often sufficient. A purely elastic analysis may be done in a single increment. In any case, at the end of each load step, the displacements, strains, and stresses are updated according to

$$\begin{aligned}
\{U(n)\} &= \{U(n-1)\} + \{\Delta U(n)\} \\
\{\varepsilon(n)\} &= \{\varepsilon(n-1)\} + \{\Delta\varepsilon(n)\} \\
\{\sigma(n)\} &= \{\sigma(n-1)\} + \{\Delta\sigma(n)\}
\end{aligned} \tag{10.4}$$

where n is the load step and $(n-1) = (0)$ means at the beginning of the first load step. For example, $\{\sigma(0)\}$ is the initial, preload stress, which may be zero. Importantly, the tangent stiffness is also updated for each element that is stressed beyond the elastic limit. Thus, the stress state in each element must be examined to determine whether the yield criterion is met; if so, then the tangent stiffness matrix $[E]$ and element stiffness matrix $[SK]$ are updated. The global stiffness matrix $[SM]$ must also be updated before application of the next load step. In a one-step elastic analysis, updating is not done; in a multi-step analysis, updating may also be avoided, provided all elements remain in the purely elastic domain. However, all elements must still be tested at the end of each load step to determine whether the yield criterion is met. In any case, in the tangent stiffness – incremental approach – to material nonlinearity, updating simply approximates integration of the differential stress–strain relations by summation.

10.2 ITERATIVE (MODIFIED NEWTON–RAPHSON) APPROACH

The Newton–Raphson approach to nonlinearity is a widely used iterative technique for coping with nonlinearity. The method is illustrated in Figure 10.1a, where the nonlinear

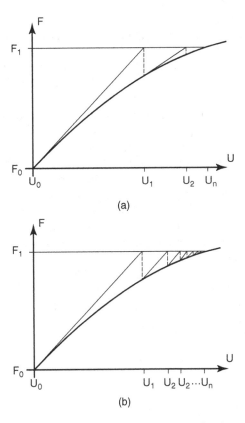

Figure 10.1 (a) Newton–Raphson and (b) modified Newton–Raphson iterations.

finite element system of equations is illustrated in analogy to a one-dimensional force–displacement curve described by $\{F(U)\} = [K(U)]\{U\}$, where the functional dependency on displacement is emphasized. The incremental form of the force–displacement curve is simply $\{\Delta F(U)\} = [K(U)]\{\Delta U\}$, which may also be obtained by expansion of the force–displacement relation in a Taylor series about a point (F_o, U_o) followed by truncation of higher-order terms than order one. Application of the load increment $\{\Delta F(1)\}$ leads to a displacement increment $\{\Delta U(1)\}$ based on the starting stiffness $[K(U_o)]$. The displacement increment is clearly too small relative to the load increment. Addition of this displacement increment to the starting displacement leads to an updated stiffness matrix $[K(U_1)]$. A new force increment $\{\Delta F(2)\}$ is then applied to obtain a second displacement increment $\{\Delta U(2)\}$ that is added to $\{U(1)\}$. The updated displacement $\{U(2)\}$ is used to compute an updated stiffness matrix $[K(U_2)]$. The updated stiffness matrix is a tangent stiffness matrix. The process is repeated until some convergence criterion is satisfied, for example when the relative change in displacement is less than 10^{-4}.

When used in conjunction with an elimination equation solving, Newton–Raphson iteration is costly because of the need to reformulate and reduce the global stiffness matrix during each iteration within each load increment. In the modified Newton–Raphson approach, shown in Figure.10.1b, the stiffness matrix is not updated after each iteration, so the original elastic stiffness matrix may be used with considerable savings in computational effort.

The main question during implementation of Newton–Raphson iteration is computation of the force increments as they change during the iteration within a load step. Such details depend on the nature of the nonlinearity. The nature of the nonlinearity may also affect convergence. A dip in the force–displacement curve will result in divergence. If the force–displacement is not monotonically decreasing, the method may also fail to converge. An alternative solution procedure is then required.

The incremental load approach with updating and the conventional Newton–Raphson iteration use a tangent stiffness. Moreover, application of the Newton–Raphson method of coping with nonlinearity is also commonly used with incremental loading. The distinction between the two approaches becomes blurred not only by terminology, but also in practice. Indeed, when incremental loading is used with a single Newton–Raphson iteration, the distinction between the two approaches is only in the equation solver, Gauss–Seidel iteration or Gaussian elimination. The former often requires more computation time for small- to medium-sized problems, but requires much less storage; the latter requires an order of magnitude more storage, but will often converge when the former may not and although fixed in solution time may be much faster than iteration when the rate of convergence is slow. Fortunately, it is not too difficult to program both and thus to have an alternative available, one solution approach should prove unsatisfactory.

Chapter 11

Time integration

Although all real physical processes take time, time does not enter explicitly into static (equilibrium) elastic and elastic–plastic analyses; strain increases concurrently with stress and immediately with application of load. Of course, the time-independent elastic–plastic material model may be oversimplified just as a purely elastic model may be an inadequate idealization for the problem at hand. If the considered material has a significant viscous component of strain, then stress will depend in some way on strain rate and time will enter explicitly into the analysis. Creep of a pillar in an underground salt mine is a situation where time-dependent inelastic analysis is needed. The "viscosity" of geologic media is generally stress-dependent, so material nonlinearity must also be taken into account. Integration in time and space is then required. Usually, the load is applied over a very short period of time relative to the time allowed for creep. In essence, the load is applied instantaneously; updating of material properties is then conveniently done at the end of a time step.

Time also enters explicitly into transient (quasi-static) analysis of seepage and in coupled problems involving porous, fractured rock masses where solid deformation and fluid flow are linked. In soil mechanics, the classic consolidation problem involves exudation of pore water from connected voids with concurrent settlement of the soil skeleton and is time-dependent. If desaturation occurs in some part of the region of interest, then material nonlinearity as well as time must be considered. In case of nonlinearity, updating of material properties (hydraulic conductivity), element stiffness matrices, and the global stiffness matrix is, again, conveniently done at the end of a time step.

In a dynamic analysis (wave propagation), inertial forces are important, and acceleration appears as the second time derivative of displacement in the equations of motion. However, transient seepage, coupled deformation and fluid flow, and creep generally involve only the first time derivative. Consider a system of equations that involve the first time derivative of unknowns $\{U\}$,

$$\{F(t)\} = [K]\{U(t)\} + [K']\{\dot{U}(t)\} \tag{11.1}$$

where the overdot denotes time differentiation and the prime distinguishes the two stiffness matrices. The solution to (11.1) first requires integration over a time interval $\Delta t = t_2 - t_1$. Thus,

$$\int_{t_1}^{t_2} \{F(t)\}\,dt - \int_{t_1}^{t_2} [K]\{U(t)\}\,dt = [K']\{\Delta U\} \tag{11.2}$$

DOI: 10.1201/9781003166283-11

which is in the standard form $\{F\} = [K']\{\Delta U\}$, ready for final solution by iteration or elimination.

A numerical approximation to the time integrals on the left-hand side of (11.2) may be obtained by assuming a linear variation over the interval Δt. Thus, for a typical function $f(t)$,

$$f(t) = (1-\theta)f(t_1) + \theta f(t_2) \tag{11.3}$$

where θ may assume a value from 0 to 1 independent of t. The time integral of (11.3) is then

$$\int_{t_1}^{t_2} f(t)\,dt = \left[(1-\theta)f(t_1) + \theta f(t_2)\right]\Delta t \tag{11.4}$$

The assumption of a linear variation implies a relatively small time step if the function under the integral sign is changing rapidly with time.

A value of $\theta = 0$ in (11.4) implies use of the initial value of the function for approximating the integral over time, a value of $\theta = 1$ uses the final value of the function, while a value of $\theta = 0.5$ uses the mean value. The first corresponds to a forward finite difference approximation; the second corresponds to a backward difference approximation, while the last is known as the Crank–Nicolson approximation.

Finite difference methods have associated stability requirements that limit the size of the time step that may be used. In this regard, the forward finite difference approximation is only conditionally stable, while the other two are unconditionally stable, thus admitting the possibility of using very large time steps. However, there may be a restriction on how small a time step can be used. Usually, this restriction arises in association with "diffusion", say, of *pore* pressure effects. The time step should generally be large enough to allow diffusion to occur across the smallest element in the mesh.

Alternatively, (11.4) can be expressed as

$$\int_{t_1}^{t_2} f(t)\,dt = f(t_1)\Delta t + \theta\left(\Delta f(t_2)\right)\Delta t \tag{11.5}$$

A geometric interpretation of (11.5) is shown in Figure 11.1. When $\theta = 0$, only the rectangular area corresponding to the first term in (11.5) is used in the integral approximation; when $\theta = 1$, a rectangular area corresponding to the final function value is used, but when $\theta = 0.5$, the triangular area corresponding to the second term in (11.5) is added that makes the result exact under the assumption of linear variation over the time increment. This observation suggests that the Crank–Nicolson approximation is more accurate and, indeed, that is the case.

The incremental form of the time integration scheme (11.5) when used in the original system (11.1) leads to a different system for the solution of the unknown change in ΔU. Thus,

$$\{F(t_1)\}\Delta t + \theta\{\Delta F\}\Delta t = [K]\{U(t_1)\}\Delta t = \left(\theta[K]\Delta t + [K']\right)\{\Delta U\} \tag{11.6}$$

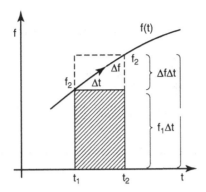

Figure 11.1 Geometric interpretation of a linear approximation to time integration.

In analysis of creep and consolidation, the total external loads {F} are often applied during a single, small time step Δt (from t_0 to t_1) and held constant thereafter. The change {ΔF} is subsequently zero. This type of loading history is shown in Figure 11.2 and approximates step function loading at time zero.

A physical example of (11.1) concerns the one-dimensional creep of a mine pillar under constant axial stress. If the material model corresponds to the Kelvin–Voigt concept of an elastic spring in parallel with a constant viscosity dashpot shown in Figure 11.3, then the differential form of the stress–strain relations is

$$\sigma(t) = E\varepsilon(t) + \eta\dot{\varepsilon}(t) \tag{11.7}$$

Comparison with (11.1) shows that E and η are analogous to $[K]$ and $[K']$, respectively, while α and ε are analogous to F and U, respectively. The solution to (11.7) under the condition of constant stress $\sigma = \sigma_o$ applied at time $t = 0$ is

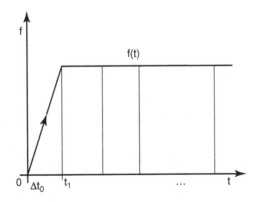

Figure 11.2 Fast ramp loading history as an approximation to step function loading at time zero.

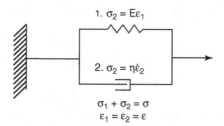

Figure 11.3 The Kelvin–Voigt viscoelastic model applied to a mine pillar.

$$\varepsilon = \left(\frac{\sigma_o}{E}\right)\left[1 - e^{-(E/\eta)t}\right]$$

(11.8)

where the stresses in the spring and the dashpot are

$$\sigma_1 = \sigma_o\left[1 - e^{-\left(\frac{E}{\eta}\right)t}\right] \text{ and } \sigma_2 = \sigma_o\left[e^{-\left(\frac{E}{\eta}\right)t}\right]$$

(11.9)

which clearly add to σ_o as they should. Figure 11.4 shows the distribution of stress in the pillar. The strain follows the stress in the spring (σ_1) and approaches the limiting value σ_o/E in time. The Kelvin–Voigt model is thus a delayed elasticity model.

Although this one-dimensional creep problem is simple to solve analytically, the exponential term often causes numerical difficulty when solving systems of equations such as (11.1), which is sometimes referred to as a "stiff" system. An example is the classic soil consolidation problem that in one dimension leads to a "diffusion" equation for the fluid pressure. Thus,

$$\frac{\partial p(z,t)}{\partial t} = c_v \frac{\partial^2 p(z,t)}{\partial z^2}$$

(11.10)

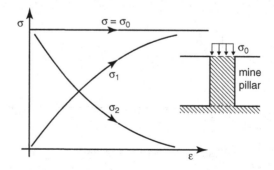

Figure 11.4 Total spring and dashpot stresses for the Kelvin–Voigt mine pillar.

where c_v is a coefficient of consolidation equal to the product of hydraulic conductivity (k) and Young's modulus (E), when Poisson's ratio is zero, with metric units of N/(m^{2}s). The solution to (11.10) is obtained by a variable separable approach and has the form

$$p(z,t) = e^{-\lambda^2 C_v t}[A\cos(\lambda z) + B\sin(\lambda z)] \tag{11.11}$$

where λ is a constant to be determined from boundary and initial conditions as are A and B which are related to no flow at the bottom of the column and zero pressure at the top at all times, and to a uniform pressure throughout the column equal to the external load at the moment the load is applied. The final result is

$$p(z,t) = p_o \left[\sum_{m=1,3...}^{m=\infty} \left(\frac{A}{m\pi} \right) \sin\left(\frac{m\pi z}{2H} \right) \exp\left[-\left(\frac{m^2\pi^2}{4H^2} \right) c_v t \right] \right] \tag{11.12}$$

which clearly shows the exponential factor that makes the problem "stiff". The "steady-state" solution which is zero pressure is approached after a long time interval. The total load is then supported entirely by the porous solid. Use of a very large time step in numerical analysis of such problems leads to a reasonable estimate of the steady-state solution.

Inspection of (11.12) suggests use of a dimensionless distance $z^* = z/H$ and a dimensionless time $t^* = c_v t/H^2$. A plot of the total stress, effective stress, and fluid pressure (compression is positive) histories for a point at the middle of the column is shown in Figure 11.5.

The displacement of the surface, that is, the settlement of the column, which is of considerable practical interest, is obtained by integrating the vertical strain over the column height. The strain follows from Hooke's law, effective stress and total stress. The result is

$$w(z,t) = \left(\frac{p_o H}{E} \right) \left[\sum_{m=1,3...}^{m=\infty} \left(\frac{8}{m^2\pi^2} \right) \cos\left(\frac{m\pi z}{2H} \right) \left(1 - \exp\left[\left(\frac{m^2\pi^2}{4H^2} \right) c_v t \right] \right) \right] \tag{11.13}$$

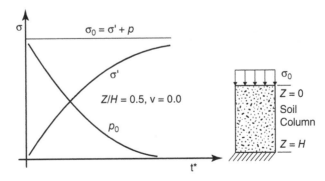

Figure 11.5 One-dimensional consolidation of a porous soil column containing an incompressible fluid (water).

where again Poisson's ratio is zero and the fluid (water) is considered incompressible in relation to the soil. Inspection of (11.13) shows that the initial displacement is zero at $t = 0+$ when the surface load is applied. If the fluid were compressible, an instantaneous elastic displacement would occur upon application of load.

The long-time settlement as t becomes indefinitely large from (11.13) is

$$w(z, \infty) = \frac{p_o}{E}(H - z) \tag{11.14}$$

which satisfies the condition of no displacement at the bottom of the column. Intuitively, one expects the maximum settlement to occur at the surface and, indeed, that is the case.

Chapter 12

Finite element seepage formulation

The finite element form of the equilibrium equations for the slow, steady flow of fluid through a rigid, porous solid ("seepage") may be obtained by application of the divergence theorem in the form of a principle of virtual power. The derivation is similar to that used to obtain the equilibrium equations of a deformable, but "dry" solid. In the case of seepage considered here, the fluid is considered "ideal", that is, incompressible and without viscosity. Flow occurs according to Darcy's law. The principle of virtual power is then

$$\int_S (\pi)(\dot{w}_n)dS + \int_V \{\dot{w}\}^T \{\gamma\}dV = \int_V \left(\frac{\partial(\pi\dot{w}_x)}{\partial x} + \frac{\partial(\pi\dot{w}_y)}{\partial y} + \frac{\partial(\pi\dot{w}_z)}{\partial z} \right)dV$$
$$+ \int_V \{\dot{w}\}^T \{\gamma\}dV \qquad (12.1)$$

where π is the fluid stress, w_n is the fluid displacement normal to the bounding surface S, the dot means differentiation with respect to time, and γ is the specific weight of the fluid flowing. Tension is considered positive, so pressure p is the negative of the fluid stress, that is, $p = -\pi$. Using the same sign convention as in elasticity maintains consistency between programs for seepage and solid mechanics. The first term on the left-hand side of (12.1) is suggestive of the rate of doing work (power) by fluid stress on fluid displacement at the surface of the considered region; the second term is the power of the body forces.

12.1 INCOMPRESSIBLE FLOW THROUGH A RIGID, POROUS SOLID

Expansion of the first term on the right-hand side of (12.1) results in

$$\int_S (\pi)(q)dS = \int_V (\pi) \left(\frac{\partial \dot{w}_x}{\partial x} + \frac{\partial \dot{w}_y}{\partial y} + \frac{\partial \dot{w}_z}{\partial z} \right) + \int_V \left(\frac{\partial \pi}{\partial x} \dot{w}_x + \frac{\partial \pi}{\partial y} \dot{w}_y + \frac{\partial \pi}{\partial z} \dot{w}_z \right)dV \quad (12.2)$$

The first integral on the right-hand side of (12.2) is the time rate of volumetric expansion of the fluid and is zero because of the assumption of incompressibility. The velocity of flow normal to the surface S is also the volumetric flow rate from the total volume V per unit area of surface and is the specific discharge q. The parentheses in (12.2) indicate scalar quantities (1×1 matrices).

DOI: 10.1201/9781003166283-12

The finite element form of (12.2) is

$$\{\pi_n\}^T \int_S [N]^T (q)dS = \{\pi_n\}^T \int_V [B]^T \{\dot{w}\})dV \tag{12.3}$$

where $\{\pi_n\}$ is an $n \times 1$ matrix of node fluid stress (n is the number of nodes per element); $[N]$ is a $1 \times n$ matrix of interpolation functions; $[B]$ is a $3 \times n$ matrix of interpolation function derivatives (node pressure-to-element pressure gradient transformation); $\{w\}$ is a 3×1 matrix of fluid displacement components; the dot denotes time derivative, as before; S is the surface of the considered element; and V is element volume. Equation (12.3) when true for arbitrary $\{\pi_n\}$ requires

$$\int_S [N]^T (q)dS = \int_V [B]^T ([k] \left\{ \begin{array}{c} \dfrac{\partial \pi}{\partial x} \\[6pt] \dfrac{\partial \pi}{\partial y} \\[6pt] \dfrac{\partial \pi}{\partial z} \end{array} \right\}) + [k]\{\gamma\})dV \tag{12.4}$$

where Darcy's law is substituted for the fluid velocities on the right-hand side of (12.3) and $[k]$ is the element hydraulic conductivity matrix. Substitution of the node pressure-to-element pressure gradient relation into (12.4) with rearrangement of terms gives

$$\int_S [N]^T (q)dS - \int_V [B]^T [k]\{\gamma\}dV = \left(\int_V [B]^T [k][B]dV \right)\{\pi_n\} \tag{12.5}$$

which in compact form is just

$$\{f\} = [k']\{\delta\} \tag{12.6}$$

where $\{f\}$, $[k']$, and $\{\delta\}$ are the element $n \times 1$ "force", $n \times n$ "stiffness", and $n \times 1$ "displacement" matrices in analogy with corresponding terms in solid mechanics. Here, fluid pressure, pressure gradients, and fluid velocity are mathematically analogous to displacements, strains, and stresses, respectively. The unknown node displacements in solid mechanics are replaced by unknown fluid stress ("pressure") in seepage analysis.

Assembly into a global system, solution for the node pressures, calculation of element pressure gradients and velocities proceed similarly. Appropriate boundary conditions analogous to specified displacements are specified node pressures. Specified flow velocity (specific discharge) normal to a boundary is analogous to a specified force at a node. In stress analysis, forces are usually specified at the surface of a body; interior nodes are usually free of externally applied forces (except for gravity loads). In seepage analysis, a flow (specific discharge) may be specified at an interior node to simulate fluid pumping or injection from a well. Of course, flow may also be specified at external boundaries.

An alternative approach is to use a hydraulic potential H as a basic unknown instead of fluid stress π. Let $H = \pi + x\gamma_x + y\gamma_y + z\gamma_z$, then the gradients of the hydraulic potential are $h_x = \partial \pi/\partial x + \gamma_x$ and similarly for h_y and h_z. Use of these relations in (12.4) results in

$$\int_S [N]^T (q) dS = \left(\int_V [B]^T [k][B] dV \right) \{H_n\} \tag{12.7}$$

where now $(H) = [N]\{H_n\}$, $\{h\} = [B]\{H_n\}$, and Darcy's law gives the seepage velocity as $v = [k]\{h\}$. The unknown $\{\delta\}$ in (12.6) is now a column matrix of node hydraulic potential values.

When the z-axis is vertical and positive down, then the hydraulic potential divided by the specific weight of fluid is just the piezometric head $\phi = \pi/\gamma + z$. According to Darcy's law, when the fluid is at rest, the pressure gradient is just balanced by the weight of fluid, that is, $\partial\pi/\partial z = -\gamma$, so $\pi = -\gamma z +$ constant. At the top of a column where $z = 0$, by choice, $\pi = 0$. Hence, $\pi = -\gamma z$, so the fluid stress is increasingly negative (a pressure) with increasing z (depth) as should be. Substitution for π into ϕ shows that ϕ is constant (zero) along a vertical line in a fluid at rest. Indeed, ϕ is constant along equipotential lines whether the fluid is at rest or in motion. In most fluid mechanics texts, pressure is considered positive and the z-axis is positive upward. Then $\phi = p/\gamma + z$. Fluid velocity according to Darcy's law is then $-[k^*]\{h\}$, where $[k^*] = [k]/\gamma$ and the minus sign is required with this alternative sign convention. When the fluid is at rest, $p = \gamma(d - z)$, where d is the height of water table above the datum from which z is measured and $(d - z)$ is simply the depth of water. Substitution of p into ϕ shows that ϕ is again constant (γd) along an equipotential line. The difference is caused by a *shift* of the datum from the top of the water table to a distance d below the water table. The result is the same for fluid pressure; the sign convention does not change the mechanics.

12.2 COMPRESSIBLE FLOW THROUGH A DEFORMABLE, POROUS SOLID

When the fluid flowing is compressible or if the solid is deformable, then the volume occupied by the fluid (per unit total volume), ζ, may change, so $\left(\partial w_x/\partial x + \partial w_y/\partial y + \partial w_z/\partial z\right)$ may change from point to point and in time. The second term on the right-hand side of (12.2) is no longer identically zero. The displacement of the fluid relative to the solid, w, is specifically, $U - u$, where U and u are fluid and solid displacements, respectively. If the fluid is incompressible, then the fluid dilatation $\left(\partial U_x/\partial x + \partial U_y/\partial y + \partial U_z/\partial z\right)$ is zero, and if the solid is incompressible or rigid, then the dilatation of the solid matrix $\left(\partial U_x/\partial x + \partial U_y/\partial y + \partial U_z/\partial z\right)$ is zero. In either case, the relative volume change of the fluid is non-zero and so may be the time rate of relative volume change of the fluid. Thus, in general, the change in fluid content per unit of time $\partial\zeta/\partial t = \partial\left(\varepsilon_v^f - \varepsilon_v^s\right)/\partial t$, where ε_v^f and ε_v^s are volumetric strains of the fluid and porous solid, respectively. Because the fluid is compressible, volumetric fluid strain is related to fluid stress: $\varepsilon_v^f = c\pi$, where c is a fluid compressibility. In the limit of incompressibility, $c = 0$. In this regard, the porous solid is continuously saturated, by assumption. If saturation is not 100%, then both air and water may be present in the connected voids of the solid. However, the concern here is only for single-phase, saturated flow.

The integrand of the second integral on the right-hand side of (12.2) is $(\pi)(\partial\zeta/\partial t) = (\pi)\left[(c)(\partial\pi/\partial t) - (\partial\varepsilon_v^s/\partial t)\right]$, which shows that the flow of fluid and deformation of the porous solid are coupled. In matrix notation, the integral assumes the form

$$\int_V (\pi)(\dot{\varsigma})dV = \int_V (\pi)\Big[(c)(\dot{\pi}) - \{c^T\}\{\dot{\varepsilon}\}\Big]dV \qquad (12.8)$$

where $\{\varepsilon\}$ is the solid strain with superscript dropped and $\{c\}^T = (1\ 1\ 1\ 0\ 0\ 0)$.

When interest is focused on seepage, the coupling is handled by assuming a relationship between fluid and solid volumetric strains, that is, by assuming $\varepsilon_v^f = f(\varepsilon_v^s)$. The simplest functional relationship is a proportionality with $\varepsilon_v^f = A\varepsilon_v^s$, where A is material constant. When this relationship is assumed, (12.8) becomes

$$\int_V (\pi)(\dot{\varsigma})dV = \int_V (\pi)(S)(\dot{\pi})dV \qquad (12.9)$$

where (S) is a "storage" coefficient. The finite element contribution from (12.9) to (12.5) is

$$\left(\int_V [N]^T (S)[N]dV\right)\{\dot{\pi}\} = [k'']\{\dot{\pi}_n\} \qquad (12.10)$$

where $[k'']$ is an element "capacitance" matrix.

The transient seepage finite element equation is then

$$\int_S [N]^T (q)dS - \int_V [B]^T [k]\{\gamma\}dV = \left(\int_V [B]^T [k][B]dV\right)\{\pi_n\}$$
$$+ \left(\int_V [N]^T (S)[N]dV\right)\{\dot{\pi}\} \qquad (12.11)$$

In more compact form, (12.11) is

$$\{f\} = [k']\{\delta\} + [k'']\{\dot{\delta}\} \qquad (12.12)$$

which certainly raises a question of time integration in view of $\{\dot{\delta}\}$.

Chapter 13

Hydro-mechanical coupling

Deformation of saturated, porous, fractured solids involves interaction between fluid flow and solid deformation, that is, "coupling", much as heat transfer and solid deformation interact through thermal stresses. Coupling is often one way in that temperature change induces thermal stresses, but the stresses do not influence heat transfer. While fluid flow influences stress, stress does not influence fluid flow, so permeability and hydraulic conductivity remain constant. Two-way coupling is possible, of course. But the analysis here assumes that coupling is just one way, embodied in the concept of "effective" stress in stress–strain relations such as Hooke's law in elasticity and hydraulic conductivity in Darcy's law, which implies a slow, laminar flow.

13.1 EFFECTIVE STRESS CONCEPT

Consider a porous medium cross section composed of a total area A equal to the sum of solid area As and fluid area Av (saturated; $v =$ voids). The sum of forces over solid and fluid areas is

$$
\begin{aligned}
F &= F_v + F_s \; \therefore \\
\sigma A &= \sigma_s A_s + \sigma_f A_f, \text{ thus,} \\
\sigma &= \sigma_s A_s / A + \sigma_f A_f / A, \text{ hence} \\
\sigma &= \sigma_s (1 - n) + \sigma_f n
\end{aligned}
\tag{13.1}
$$

where n is the area porosity. The area porosity is often assumed to be equal to the volume porosity.

The last of (13.1) can be written in terms of *effective solid and fluid stresses*, both per unit total area. Thus,

$$
\sigma = \sigma' + \pi'
\tag{13.2}
$$

The advantage of (13.2) over the more conventional $\sigma = \sigma' - p$, where p is the pore pressure, is consistency. Both stresses are reckoned per unit total area, so all forces are obtained by simply multiplying by total area. Both stresses are also reckoned by the same sign conventions, say tension is positive as usual in finite element codes.

DOI: 10.1201/9781003166283-13

13.2 FINITE ELEMENT FORMULATION

In matrix form using Hooke's law, one has

$$\{\sigma\}_{6\times1} = [E]_{6\times6}\{\varepsilon\}_{6\times1} + \{c\}_{6\times1}(\pi)_{1\times1} \tag{13.3}$$

where the column matrix $\{c\}$ often has rows (1 1 1 0 0 0), so $\{c\}(\pi)$ is $\{\pi\ \pi\ \pi\ 0\ 0\ 0\}6\times1$. This matrix is also called a coupling matrix and may have values different from 1s and 0s. Recall that effective shear stresses are just the shear stresses and no fluid stress is involved.

Recall that the change in fluid content per unit volume is designated with the symbol ζ. This property is simply the change in fluid volume relative to the solid volume change. Thus, $\zeta = \varepsilon_v^f - \varepsilon_v^s$ and

$$\zeta = c\pi - \{c\}^t\{\varepsilon\} \tag{13.4}$$

where c is the fluid compressibility, the reciprocal of the fluid bulk modulus, and $\{c\}$ is the matrix introduced in (13.3).

Equations (13.3) and (13.4) constitute Biot's law for poroelastic behavior, that in matrix form is

$$\begin{Bmatrix} \{\sigma\} \\ -\varsigma \end{Bmatrix}_{7\times1} = \begin{bmatrix} [E] & \{c\} \\ \{c\}^t & -c \end{bmatrix}_{7\times7} \begin{Bmatrix} \{\varepsilon\} \\ \pi \end{Bmatrix}_{7\times1} \tag{13.5}$$

Interestingly enough, the 7×7 material properties matrix in (13.5) is symmetric.

Application of the divergence theorem to total stress and to fluid stress leads to a continuum result suitable for finite element approximation. In case of total stress,

$$\int_S \{u\}^t\{T\}\,dS + \int_V \{u\}^t\{\gamma\}\,dV = \int_V \{\varepsilon\}^t\{\sigma\}\,dV \tag{13.6}$$

which is in the same form as that for the dry case. The left-hand side (LHS) of (13.6) in finite element form is

$$\text{LHS (fem)} = \{\delta\}^t\left(\int_S [N]^t\{T\}\,dS + \int_V [N]^t\{\gamma\}\,dV\right) \tag{13.7}$$

where the integrals in parentheses are node forces as before. The right-hand side (RHS) is

$$\text{RHS (fem)} = \{\delta\}^t\left(\int_V [B]^t(\{\sigma'\} + \{c\}\pi\,dV\right) \tag{13.8}$$

where substitution for total stress is made using (13.2). Further substitution gives

$$\text{RHS (fem)} = \{\delta\}^t\left(\int_V \left([B]^t[E][B]\{\delta\} + [B]^t\{c\}[N_p]\{\pi_n\}\right)dV\right)$$

where the subscript p denotes "fluid". After combining LHS and RHS, one obtains for an element the stiffness equations:

$$\{f\} = [k]\{\delta\} + [k']\{\pi_n\} \tag{13.7}$$

Assembly would lead to the master stiffness equations in the usual manner. Dimensions of the matrices depend on the number of element nodes, of course. For example, an 8-node linear displacement brick with three unknown displacements at each node would lead to $[N]$, a 3×24 matrix, and $[B]$, a 6×24 matrix, with $[E]$, a 6×6 matrix. If reduced in view of plane stress, plane strain, or axial symmetry, then $[E]$ would be 3×3 or more likely 4×4 to cover all two-dimensional options. Other matrices would be reduced according to element node numbers, of course.

The fluid moves relative to the solid, so additional work is done and power is needed. In rate form, one has $\int_S \{\dot{u}\}^t \{T\} dS + \int_V \{\dot{u}\}^t \{\gamma\} dV = \text{LHS}$, where the dot denotes time derivative. Thus,

$$\int_S \begin{Bmatrix} \pi n_x \\ \pi n_y \\ \pi n_z \end{Bmatrix}^t \begin{Bmatrix} \dot{w}_x \\ \dot{w}_y \\ \dot{w}_z \end{Bmatrix} dS + \int_V \{\dot{w}\}^t \{\gamma\} dV = \text{LHS} \tag{13.8}$$

where substitutions for fluid stress and velocity are made. Application of the divergence theorem to the surface integral gives RHS. Thus,

$$\int_V \pi\left(\left(\partial\dot{w}_x/\partial x + \partial\dot{w}_y/\partial y + \partial\dot{w}_z/\partial z\right) + \left(\dot{w}_x\,\partial\pi/\partial x + \dot{w}_y\,\partial\pi/\partial y + \dot{w}_z\,\partial\pi/\partial z\right)\right) dV$$
$$+ \int_V \{\dot{w}\}^t \{\gamma^f\} dV = \text{RHS} \tag{13.9}$$

Because the volume integral appears on both sides, it may be dropped from further consideration. Equating RHS to LHS gives

$$\int_S \pi\dot{w}_n dS = \int_V \left(\pi\left(\partial\dot{w}_x/\partial x + \partial\dot{w}_y/\partial y + \partial\dot{w}_z/\partial z\right) + \left(\dot{w}_x\,\partial\pi/\partial x + \dot{w}_y\,\partial\pi/\partial y + \dot{w}_z\,\partial\pi/\partial z\right)\right) dV \tag{13.10}$$

where dw_n/dt is fluid flow relative to solid motion normal to the surface S. This flow is also the specific discharge, flow rate per unit of surface area, q. In finite element form, the LHS is

$$\text{LHS} = \{\pi_n\} \int_S [N_p]^t (q) dS \tag{13.11}$$

The RHS requires more work in several steps. The first step involves some evident substitutions:

$$\text{RHS} = \{\pi_n\} \int_V [N_p]^t \left(\partial\left(\varepsilon_v^f - \varepsilon_v^s\right)/\partial t\right) + [B_p]^t \begin{Bmatrix} \dot{w}_x \\ \dot{w}_y \\ \dot{w}_z \end{Bmatrix} dV \tag{13.12}$$

where the subscript p denotes fluid matrix. The next step involves more substitutions.

$$\text{RHS} = \{\pi_n\}\int_V \left[N_p\right]^t \left(c\dot{\pi} - \{c\}^t\left[B_p\right]^t[k]\left(\begin{Bmatrix} \partial\pi/\partial x \\ \partial\pi/\partial x_y \\ \partial\pi/\partial x_z \end{Bmatrix} + \{\gamma\}\right)\right)dV \qquad (13.13)$$

Continuing,

$$\text{RHS} = \{\pi_n\}\int_V \left[N_p\right]^t \left(c\dot{\pi} - \{c\}^t[B]\{\dot{\delta}\}\right) + \left[B_p\right]^t[k]\left(\left[B_p\right]\{\pi_n\} + \{\gamma\}\right)dV \qquad (13.14)$$

Both sides are multiplied by the fluid node pressure and so must be equal. Thus,

$$\int_S \left[N_p\right]^t(q)\,dS = \int_V \left[N_p\right]^t \left(c\dot{\pi} - \{c\}^t[B]\{\dot{\delta}\}\right) + \left[B_p\right]^t[k]\left(\left[B_p\right]\{\pi_n\} + \{\gamma\}\right)dV \quad (13.15)$$

In condensed form,

$$\int_S [N_p]^t(q)\,dS + \int_V [B_p]^t[k]\{\gamma^f\}\,dV = [k'_{\pi\pi}]\{\dot{\pi}\} - [k_{\pi\delta}]\{\dot{\delta}\} + [k_{\pi\pi}]\{\pi_n\} \qquad (13.16)$$

Finally, the fluid portion of the element stiffness equation is

$$\{f_\pi\} = [k'_{\pi\pi}]\{\dot{\pi}\} - [k_{\pi\delta}]\{\dot{\delta}\} + [k_{\pi\pi}]\{\pi_n\} \qquad (13.17)$$

In the above, $[k]$ without subscripts is the hydraulic conductivity matrix that involves permeability, a material property, viscosity, and specific weight of the fluid flowing. Altogether, for the coupled hydro-mechanical problem in poroelasticity, one has

$$\begin{aligned} \{f\} &= [k_{\delta\delta}]\{\delta\} + [k_{\delta\pi}]\{\pi_n\} \\ \{f_\pi\} &= [k'_{\pi\pi}]\{\dot{\pi}\} - [k_{\pi\delta}]\{\dot{\delta}\} + [k_{\pi\pi}]\{\pi_n\} \end{aligned} \qquad (13.19)$$

where subscripts have been added for clarity.

An interesting question arises after inspecting (13.19) that concerns symmetry and whether $[k_{\delta\pi}] = [k_{\pi\delta}]$, that is, whether $[N_p]^t\{c\}^t[B] = \left([B]^t\{c\}[N_p]\right)^t$. Inspection shows the equality to hold. In this regard, the system in (13.19) can be put into a symmetric form and thus enable equation solvers that assume symmetry. Thus,

$$\begin{aligned} \{f\} &= [k_{\delta\delta}]\{\delta\} + [k_{\delta\pi}]\{\pi_n\} \\ -\left(\{f_\pi\} - [k_{\pi\pi}]\{\pi_n\}\right) &= [k_{\pi\delta}]\{\dot{\delta}\} - [k'_{\pi\pi}]\{\dot{\pi}_n\} \end{aligned} \qquad (13.20)$$

The time derivatives in (13.20) require integration, so one can follow the evolution of solid deformation and fluid flow with the passage of time from application of initial conditions to steady-state conditions when changes in time become negligibly small. After time integration of the second of (13.20), one obtains

$$\begin{aligned} \{f\} &= [k_{\delta\delta}]\{\delta\} + [k_{\delta\pi}]\{\pi_n\} \\ -\int_{\Delta t} \left(\{f_\pi\} - [k_{\pi\pi}]\{\pi_n\}\right)dt &= [k_{\pi\delta}]\{\delta\} - [k'_{\pi\pi}]\{\pi_n\} \end{aligned} \qquad (13.21)$$

where integration is over a time interval Δt. The form (13.21) is symmetric as inspection of the RHS shows. Each sub-matrix is symmetric. In a more compact notation, (13.21) is

$$\begin{Bmatrix} \{f_\delta\} \\ \{f_\pi\} \end{Bmatrix} = \begin{bmatrix} [k_{\delta\delta}] & [k_{\delta\pi}] \\ [k_{\pi\delta}] & -[k_{\pi\pi}] \end{bmatrix} \begin{Bmatrix} \{\delta\} \\ \{\pi_n\} \end{Bmatrix} \tag{13.22}$$

In global, assembled form (13.13) is compactly stated as

$$\{F\} = [K]\{\Delta\} \tag{13.23}$$

where now the "displacement" Δ at each node is composed of three components of a displacement vector (u, v, w) and node pressure π; each node force is also a four-component vector.

Equation (13.23) is solved for each time step that are relatively small to begin with, but are increased as time passes. Often, the time step is increased rather rapidly. In this work, the time step is specified in the run stream and several run streams are needed to reach steady-state equilibrium conditions. By using several run streams in sequence with increasing time steps, output at each time step may be obtained until steady state is closely approached.

Chapter 14

Boundary element formulations

The boundary element method (BEM) is the second most popular numerical method for engineering analyses after the finite element method (FEM). The method has applications in heat transfer, fluid flow, acoustics, and other physical phenomena, although the focus here is on geomechanics. There are advantages with both methods. The main advantage of BEM is the reduced dimensionality of the problem because only the surfaces of excavations need to be discretized. However, BEM leads to a fully populated master stiffness matrix, which leads to problem size limitations, unlike FEM that leads to a sparsely populated master stiffness matrix. Discretizing only the surface of an opening makes mesh construction of complex shapes much easier. In case of simple shapes such as a circular hole, the advantage is not so evident because of the ease of meshing in either approach. In case of a network of underground mine drifts, crosscuts, raises, and stopes, the advantage of the BEM is much more evident. In cases of civil works such as intersections of tunnels of different sizes, the advantage of BEM is again evident. Two major disadvantages of BEM are (1) coping with different rock types and (2) deformation beyond the elastic limit. In rock mechanics, BEM is used mainly for linearly elastic, homogeneous material analyses.

The mix of physical motivation and mathematical rigor that is used by Starfield and Crouch (1983) to formulate various BEM models is followed here and is intended as a start of exploration of BEM, which has advanced enormously through the intervening years. In this regard, one notes that there are two basic formulations of BEM, an "indirect" formulation and a "direct" formulation.

14.1 INDIRECT FORMULATION

The boundary element method (BEM)in geomechanics makes use of singular solutions in the mathematical theory of elasticity. Of particular importance is the Kelvin solution to the problem of a force applied at a point in an infinite body. In the case of plane strain the solution the force F is a line load along the z-axis. Formulas for displacements are (Crouch and Starfield 1983)

$$u = \left(F_x/2G\right)\left[(3-4v)g - x\left(\partial g/\partial x\right)\right] + \left(F_y/2G\right)\left[-y\left(\partial g/\partial x\right)\right]$$
$$v = \left(F_x/2G\right)\left[-x\left(\partial g/\partial y\right)\right] + \left(F_y/2G\right)\left[(3-4v)g - \left[-y\left(\partial g/\partial y\right)\right]\right]$$
$$g(x,y) = \left[-1/4\pi(1-v)\right]\left[\ln\sqrt{(x^2+y^2)}\right]$$

(14.1)

DOI: 10.1201/9781003166283-14

and for the stresses

$$\sigma_{xx} = F_x\left[2(1-v)\partial g/\partial x - x\left(\partial^2 g/\partial^2 x\right)\right] + F_y\left[2v\partial g/\partial y - y\left(\partial^2 g/\partial^2 x\right)\right]$$
$$\sigma_{yy} = F_x\left[2v\partial g/\partial x - x\left(\partial^2 g/\partial^2 y\right)\right] + F_y\left[2(1-v)\partial g/\partial y - y\left(\partial^2 g/\partial^2 y\right)\right]$$
$$\tau_{xy} = F_x\left[(1-2v)\partial g/\partial y - x\left(\partial^2 g/\partial x\partial y\right)\right] + F_y\left[(1-2v)\partial g/\partial x - y\left(\partial^2 g/\partial x\partial y\right)\right]$$
$$\sigma_{zz} = v\left(\sigma_{xx} + \sigma_{yy}\right), \quad \tau_{zx} = \tau_{zy} = 0 \tag{14.2}$$

in consideration of plane strain conditions and material isotropy.

In the case of constant tractions P_x and P_y applied to a line segment as illustrated in Figure 14.1, point forces are $F_i(\eta) = P_i d\eta$. Results for displacements and stresses according to Crouch and Starfield (1983) are

$$u = (P_x/2G)\left[(3-4v)f - x(\partial f/\partial x)\right] + (P_y/2G)\left[-y(\partial f/\partial x)\right]$$
$$v = (P_x/2G)\left[-x(\partial f/\partial y)\right] + (P_y/2G)\left[(3-4v)f - y(\partial f/\partial y)\right]$$
$$f(x,y) = \int_{-a}^{+a} g(x-\eta, y)d\eta \tag{14.3}$$

$$\sigma_{xx} = P_x\left[2(1-v)\partial f/\partial x - x\left(\partial^2 f/\partial^2 x\right)\right] + P_y\left[2v\partial f/\partial y - y\left(\partial^2 f/\partial^2 x\right)\right]$$
$$\sigma_{yy} = P_x\left[2v\partial f/\partial x - x\left(\partial^2 f/\partial^2 y\right)\right] + P_y\left[2(1-v)\partial f/\partial y - y\left(\partial^2 f/\partial^2 y\right)\right]$$
$$\tau_{xy} = P_x\left[(1-2v)\partial f/\partial y - x\left(\partial^2 f/\partial x\partial y\right)\right] + P_y\left[(1-2v)\partial f/\partial x - y\left(\partial^2 f/\partial x\partial y\right)\right]$$
$$\sigma_{zz} = v(\sigma_{xx} + \sigma_{yy}), \quad \tau_{zx} = \tau_{zy} = 0 \tag{14.4}$$

The function $f(x,y)$ is

$$f(x,y) = \left(\frac{-1}{4\pi(1-v)}\right)\left\{y\left[\arctan\left(\frac{y}{x-a}\right) - \arctan\left(\frac{y}{x+a}\right)\right]\right.$$
$$\left. - (x-a)\ln\sqrt{(x-a)^2 + y^2} + (x+a)\ln\sqrt{(x+a)^2 + y^2}\right\} \tag{14.5}$$

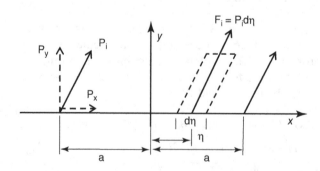

Figure 14.1 Constant tractions and forces distributed over a line segment $(-a, +a)$.

An excavation cross section may be approximated by a series of small line segments as shown in Figure 14.2. Inside the curve is the interior; outside is the Exterior. Figure 14.2 shows the numerical model where loads P_n and P_s are applied.

Superposition of loads P_n^j and P_s^j after suitable transformation of coordinates to a common system leads to the system

$$
\begin{aligned}
\tau_n^i &= \sum_{j=1}^{j=n} A_{ss}^{ij} P_s^j + \sum_{j=1}^{j=n} A_{sn}^{ij} P_n^j \\
&\qquad\qquad\qquad\qquad (i = 1, n) \\
\sigma_n^i &= \sum_{j=1}^{j=n} A_{ns}^{ij} P_s^j + \sum_{j=1}^{j=n} A_{nn}^{ij} P_n^j
\end{aligned}
\tag{14.6}
$$

The terms A_{ss}^{ij}, A_{ns}^{ij}, A_{sn}^{ij}, and A_{nn}^{ij} are *influence coefficients* that relate the load on the jth segment to the stress on the ith segment. The effects of loads on a segment are not confined to the segment, but influence loads at all other segments. In this regard, the boundary loads P_n and P_s are *fictitious* stresses, not the actual boundary stresses σ_n and τ_n. Indeed, this formulation of BEM is known as the fictitious stress formulation, an *indirect* formulation. The system of equations (14.6) can be expressed in matrix form. Thus, with the matrix dimensions as subscripts outside braces and brackets,

$$
\left\{ \begin{matrix} \{\tau\} \\ \{\sigma\} \end{matrix} \right\}_{2n\times1} = \left[\begin{matrix} [A_{ss}][A_{sn}] \\ [A_{ns}][A_{nn}] \end{matrix} \right]_{2n\times2n} \left\{ \begin{matrix} \{P_s\} \\ \{P_n\} \end{matrix} \right\}_{2n\times1}
\tag{14.6a}
$$

Solution to 14.5 applies to both problem domains and is readily done by elimination or iteration when the system is relatively large. An example of an interior problem is solution to a "point-loaded" disk commonly used to determine tensile strength of rock. A common exterior problem concerns stress concentration at the wall of a circular hole (shaft; tunnel). Details are given in Crouch and Starfield (1983).

Another indirect formulation is one based on a problem involving a discontinuity in displacement across a straight line segment located in an infinite elastic solid. This

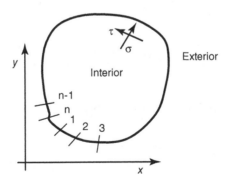

Figure 14.2 Interior and exterior physical domains bounded by a closed curve approximated by n line segments subject to boundary normal and shear stresses σ and τ.

method was pioneered by Crouch (1976) and begins with the definition of a displacement discontinuity $D_x = u(x,-0) - u(x,+0)$, $D_y = v(x,-0) - v(x,+0)$, where u and v are x and y components of displacement at a line segment $(-a, +a)$ on the x-axis. According to Crouch and Starfield, the problem has the solution for displacements

$$u = D_x\left[2(1-v)\partial f/\partial y - y\partial^2 f/\partial^2 x\right] + D_y\left[-(1-2v)\partial f/\partial x - y\partial^2 f/\partial x\partial y\right]$$
$$v = D_x\left[(1-2v)\partial f/\partial x - y\partial^2 f/\partial x\partial y\right] + D_y\left[2(1-v)\partial f/\partial y - y\partial^2 f/\partial^2 y\right] \qquad (14.7)$$

and for stresses

$$\sigma_{xx} = 2GD_x\left[2\partial^2 f/\partial y\partial x + y\partial^3 f/\partial x\partial^2 y\right] + 2GD_y\left[\partial^2 f/\partial^2 y + y\partial^3 f/\partial^3 y\right]$$
$$\sigma_{yy} = 2GD_x\left[-y\partial^3 f/\partial x\partial^2 y\right] + 2GD_y\left[\partial^2 f/\partial^2 y - y\partial^3 f/\partial^3 y\right]$$
$$\tau_{xy} = 2GD_x\left[\partial^2 f/\partial^2 y + y\partial^3 f/\partial^3 y\right] + 2GD_y\left[-y\partial^3 f/\partial x\partial^2 y\right]$$
$$\sigma_{zz} = v\left(\sigma_{xx} + \sigma_{yy}\right), \quad \tau_{zx} = \tau_{zy} = 0 \qquad (14.8)$$

The function $f(x,y)$ is

$$f(x,y) = \left[\frac{-1}{4\pi(1-v)}\right]\left\{y\left[\arctan\left(\frac{y}{x-a}\right) - \arctan\left(\frac{y}{x+a}\right)\right]\right.$$
$$\left. -(x-a)\ln\sqrt{(x-a)^2 + y^2} + (x+a)\ln\sqrt{(x-a)^2 + y^2}\right\} \qquad (14.9)$$

After evaluating derivatives, stresses along the x-axis

$$\sigma_{xx}(x,0) = \left[\frac{-aG}{\pi(1-v)}\right]D_y\left(\frac{1}{x^2 - a^2}\right)$$
$$\sigma_{yy}(x,0) = \left[\frac{-aG}{\pi(1-v)}\right]D_y\left(\frac{1}{x^2 - a^2}\right)$$
$$\tau_{xy}(x,0) = \left[\frac{-aG}{\pi(1-v)}\right]D_x\left(\frac{1}{x^2 - a^2}\right)$$
$$\sigma_{zz} = v(\sigma_{xx} + \sigma_{yy}), \quad \tau_{zx} = \tau_{zy} = 0 \qquad (14.10)$$

If a line segment of length $2a$ is centered at $x = x^i$, $y = 0$, then the normal stress at the midpoint of the ith element caused by a displacement discontinuity at the jth element is

$$\sigma_{yy}(x^i,0) = \left[\frac{-a^jG}{\pi(1-v)}\right]D_y^j\left[\frac{1}{(x^i - x^j)^2 - (a^j)^2}\right] \qquad (14.11)$$

where the discontinuity over the element segment $\left|x - x^j\right| \le a^j, y = 0$ is D_y^j. Superposition leads to

$$\sigma_{yy}^i = \sigma_{yy}(x^i,0) = \sum_{j=1}^{j=n} A^{ij}D_y^j \qquad (i = 1,n) \qquad (14.12)$$

where the influence coefficients $A^{ij} = \left[\dfrac{-a^j G}{\pi(1-v)} \right] \left[\dfrac{1}{(x^i - x^j)^2 - (a^j)^2} \right]$

The $n \times n$ system 14.12 has the matrix form

$$\{\sigma(x,0)\}_{n\times 1} = [A]_{n\times n}\{D_y\}_{n\times 1} \qquad (14.12a)$$

and is readily solved for the unknowns D_y^j by elimination or iteration. Back-substitution in (14.10) gives the stresses at the boundary. Displacements can also be found.

When the line segments define a domain boundary as in Figure 14.3, coordinate transformations are required to place variables in a common system. The result is a system similar in form to (14.6). Thus,

$$\tau_n^i = \sum_{j=1}^{j=n} A_{ss}^{ij} D_s^j + \sum_{j=1}^{j=n} A_{sn}^{ij} D_n^j$$
$$\hspace{4cm} (i = 1, n) \qquad (14.13)$$
$$\sigma_n^i = \sum_{j=1}^{j=n} A_{ns}^{ij} D_s^j + \sum_{j=1}^{j=n} A_{nn}^{ij} D_n^j$$

In matrix form,

$$\begin{Bmatrix} \{\tau\} \\ \{\sigma\} \end{Bmatrix}_{2n\times 1} = \begin{bmatrix} [A_{ss}][A_{sn}] \\ [A_{ns}][A_{nn}] \end{bmatrix}_{2n\times 2n} \begin{Bmatrix} \{D_s\} \\ \{D_n\} \end{Bmatrix}_{2n\times 1} \qquad (14.13a)$$

After solution for D_s^j and D_n^j, back-substitution in (14.10) allows for determination of the boundary stresses.

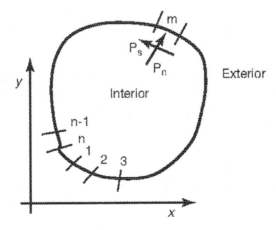

Figure 14.3 Numerical model domains bounded by a closed curve approximated by n boundary elements subject to boundary normal and shear loads P_n and P_s.

A system for displacements follows a similar development and leads to a similar system and eventual determination of boundary displacements. Details are given in Crouch and Starfield (1983).

Calculation of stress away from a boundary requires additional work in both formulations that were outlined in the preceding text. The procedure leading to the influence coefficients for points away from a boundary is much the same as for boundary points and results in a system similar to (14.13). The resulting system must be solved for each interior point.

A special procedure is also developed for the normal stress acting at the boundary curve. This stress is the one of most interest in almost all analyses. The normal and shear stresses at an excavation boundary are usually zero. A pressurized cavity is an exception. As a reminder, BEM analyses result in stress changes induced by excavation. The final stress distribution is obtained by adding these stress changes to the stresses that existed prior to excavation, that is, the initial or in situ stresses.

14.2   DIRECT FORMULATION

The direct formulation of BEM is based on a reciprocal theorem in the mathematical theory of elasticity. This theorem considers two boundary value problem solutions and states that the work done by the stresses of the first solution $\left(\tau^1, \sigma^1\right)$ on the displacement of the second solution $\left(u^1, v^1\right)$ is equal to the work done by the stresses of the second solution $\left(\tau^2, \sigma^2\right)$ on the displacement of the first solution $\left(\tau^2, \sigma^2\right)$. The superscripts indicate the first and second solutions and are not exponents. The theorem is

$$\int_C \left(\tau^1 v^2 + \sigma^1 u^2\right) ds = \int_C \left(\tau^2 v^1 + \sigma^2 u^1\right) ds \tag{14.14}$$

where ds is a differential of arc length along the contour C. After approximating C by a series of segments Δs_j and assuming the stresses are constant over a segment, the result is

$$\sum_{j=1}^{j=n} \tau^j \int_{j\Delta s} v^2\, ds + \sum_{j=1}^{j=n} \sigma^j \int_{j\Delta s} u^2\, ds = \sum_{j=1}^{j=n} v^j \int_{j\Delta s} \tau^2\, ds + \sum_{j=1}^{j=n} u^j \int_{j\Delta s} \tau^2\, ds \tag{14.15}$$

where the quantities with superscript j are evaluated at the midpoint of jth segment $j\Delta s$. After considering the second solution as known from application of point forces to the midpoints of the segments approximating the contour C, the system takes on the influence function form

$$\begin{aligned}
\sum_{j=1}^{j=n} B_{ss}^{ij} \tau^j + \sum_{j=1}^{j=n} B_{sn}^{ij} \sigma^j &= \sum_{j=1}^{j=n} A_{ss}^{ij} v^j + \sum_{j=1}^{j=n} A_{sn}^{ij} u^j \\
\sum_{j=1}^{j=n} B_{ns}^{ij} \tau^j + \sum_{j=1}^{j=n} B_{nn}^{ij} \sigma^j &= \sum_{j=1}^{j=n} A_{ns}^{ij} v^j + \sum_{j=1}^{j=n} A_{nn}^{ij} u^j
\end{aligned} \qquad \left(i = 1, n\right) \tag{14.15}$$

according to Crouch and Starfield (1983) where details are presented. Additional development brings (14.15) into the form

$$
\begin{aligned}
Y_s^i &= \sum_{j=1}^{j=n} C_{ss}^{ij} X_s^j + \sum_{j=1}^{j=n} C_{sn}^{ij} X_n^j \\
Y_n^i &= \sum_{j=1}^{j=n} C_{ns}^{ij} X_s^j + \sum_{j=1}^{j=n} C_{nn}^{ij} X_n^j
\end{aligned}
\qquad (i = 1, n)
\tag{14.16}
$$

In matrix notation,

$$
\left\{ \begin{matrix} \{Y_s\} \\ \{Y_n\} \end{matrix} \right\}_{2n \times 1} = \left[\begin{matrix} [C_{ss}][C_{sn}] \\ [C_{ns}][C_{nn}] \end{matrix} \right]_{2n \times 2n} \left\{ \begin{matrix} \{X_s\} \\ \{X_n\} \end{matrix} \right\}_{2n \times 1}
$$

The influence coefficients C_{ss}^{ij}, C_{sn}^{ij}, C_{ns}^{ij}, and C_{nn}^{ij} are known quantities; the left-hand-side quantities Y_s^i and Y_n^i are also known from prescribed boundary conditions. The unknowns are the X_s^i and X_n^i that are also combinations of boundary quantities. Solution to the system of equations (14.16) allows for determination of the unknowns in X_s^i and X_n^i.

An example problem demonstrating the use of a boundary element problem involves twin circular tunnels. A solution to the problem is given by Ling (1948). Stress concentrations at the top, bottom, outside, and inside of the circular sections are given by Pariseau (2007). The results are illustrated in Figure 14.3. These results were obtained using a Rocscience[1] boundary element program (EXAMINE2D). These results scarcely begin to indicate what can be obtained by a BEM program. Commercial programs such as the one used here have many tools for color plotting and graphing results in many different ways, including strength factors when a failure criterion is available in analyses. Table 14.1 is a brief comparison of stress concentrations from theory and BEM at critical points on the excavation boundaries. The uniaxial vertical load is not a realistic load and is used only for demonstration purposes here. BEM values in the table are also only approximate because they depend on a mouse pick location, although they do indicate a reasonable agreement. Computation time is only a second or so (Figure 14.4).

Another application of BEM based on the displacement discontinuity formulation relates to underground mining of thin ore bodies of great lateral extent. Several

Table 14.1 Stress concentration at the outside (A), inside (B), and bottom (C) of the tunnels

Stress concentration	A (Sig 1)	B (Sig 1)	C (Sig 3)
Theory	3.07	3.02	−0.81
BEM	3.06	2.83	−0.88

1 Rocscience, Inc. 54 St. Patrick Street, Toronto, Ontario, Canada, M5T 1V1.

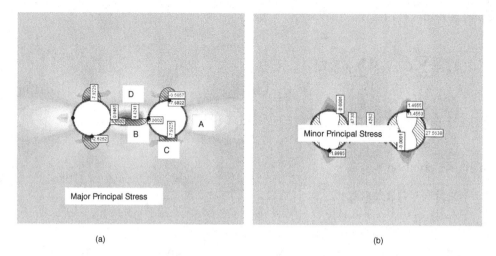

(a)   (b)

Figure 14.4 Grayscale plots of the major and minor principal stresses about twin circular
tunnels separated by a distance equal to tunnel diameter. Loading is 9 units
in the vertical direction only. Analysis is plane strain and purely elastic. Compression is positive.

program packages are available for addressing related problems of stress and displacements induced by mining. One program of interest is MULSIM/NL (Zipf 1992)
that has evolved from the displacement discontinuity formulation advanced by Sinha
(1979); another in the mining community is LAMODEL (Heasley 1998). There are
many others. Figure 14.5 shows a possible mining layout for analysis by a program
such as MULSIM/NL, LAMODEL, or similar programs.

In analysis of the problem by these programs, **the pillars are not in the solution
domain**, and for this reason, such BEM programs are not suitable for pillar design.
In practice, effects of pillars on the material above and below the mining horizon
arise from specified boundary stresses. In some instances, even the extent of pillar
failure is specified at the top and bottom surfaces. But claims for pillar design using
these programs are circular because distributions of stress at pillar tops and bottoms

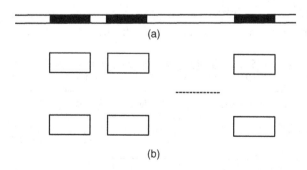

Figure 14.5 Mine layout for BEM analysis using the displacement discontinuity formulation. Rectangles are mine pillars: (a) vertical section and (b) plan view.

are specified in advance (as input boundary conditions). While distributions of stress in pillars are necessary for engineering evaluation of pillar safety, using such BEM programs for pillar design is ill-advised.

REFERENCES

Banerjee, P.K. and R. Butterfield (1981) *Boundary Element Methods in Engineering Science.* McGraw Hill, New York, pp 452.

Beer, G. (2001) *Programming the Boundary Element Method: An Introduction for Engineers.* John Wiley & Sons, New York, pp 457.

Crouch, S.L. (1976) "Solution of a Plane Elasticity Problems by the Displacement Discontinuity Method." *Int. J. Num. Methods Eng.*, Vol. 10, pp 301–343.

Crouch, S.L. and A.M. Starfield (1983) *Boundary Element Methods in Solid Mechanics.* George Allen & Unwin, London, pp 322.

Heasley, K.A. (1998) "Numerical Modeling of Coal Mines with a Laminated Displacement Discontinuity Code." Ph.D. Dissertation, Colorado School of Mines.

Ling, C.-H. (1948) "On the Stresses in a Plate Containing Two Circular holes". *J. Appl. Phys.*, Vol. 19, pp 77–82.

Pariseau, W.G (2007) *Design Analysis in Rock Mechanics.* Taylor & Francis, London, pp 560.

Sinha, P.K. (1979) "Displacement Discontinuity Technique for Analyzing Stresses and Displacements due to Mining in Seam Deposits." Ph.D. thesis, U. of Minnesota.

Telles, J.C.F. (1983) *The Boundary Element Method Applied to Inelastic Problems.* Springer-Verlag, Berlin, pp 243.

Zipf, R.K., Jr (1992) "MULSIM/NL Theoretical and Programmer's Manuel." *U.S. Bureau of Mines Information Circular* 9321.

Chapter 15

Distinct element formulations

The distinct element method (DEM), also known as the discrete element method, was developed originally by Cundall (1971, 1974). Subsequent development by Cundall et al. (1978) and Cundall and Hart (1985) led to the popular Itasca Consulting Group, Inc. programs UDEC and 3DEC (two- and three-dimensional distinct element programs). DEM is a numerical procedure for analysis of blocky rock systems. Elements are rock blocks defined by rock joint geometry. The objective of DEM is to take rock joints into account in the determination of rock block motions that collectively define the motion of a jointed rock mass system subject to forces of gravity, contact forces between adjacent rock blocks, and externally applied forces.

Subsequent development of DEM by Cundall and Strack (1978, 1979) considered *particles* (disks and spheres) as elements for application to materials handling problems in programs BALL (two-dimensional) and TRUBAL (three-dimensional). DEM is now in two forms: block DEM and particle DEM. Jing and Stephansson (2007) described DEM in detail; O'Sullivan (2011) offered a well-read review of particle DEM. Elements in block DEM may now be polygons and polyhedrons in two- and three-dimensional systems, respectively, as well as disks and spheres. Block DEM elements were originally rigid bodies, but may be deformable. A third hybrid form using clusters of bonded particles has also been developed in two- and three-dimensional programs, e.g., PFC2D and PFC3D (Itasca Consulting Group), "particle flow codes", not to be confused with computational fluid dynamic programs. Particles in these particle flow programs may be polygons, disks, polyhedrons, or spheres and may fracture and fail during an analysis. Indeed, one of the original DEM programs used polygonal blocks in a materials handling application illustrated in Figure 15.1.

Description of block and particle motions is governed by (1) physical laws, (2) kinematics, (3) material laws when blocks are deformable, and (4) *contact mechanics*. The last, contact mechanics, refers to forces of interaction between neighboring blocks and is at the core of DEM analyses.

Discontinuous deformation analysis (DDA) is another numerical method for describing the motion of blocky or particulate rock systems and was first formulated by Shi and Goodman (1984, 1985). Details are given in Shi (1988). While forces are the focus of DEM, displacements are emphasized in DDA. However, the core of both formulations is the mechanics of contacts.

DOI: 10.1201/9781003166283-15

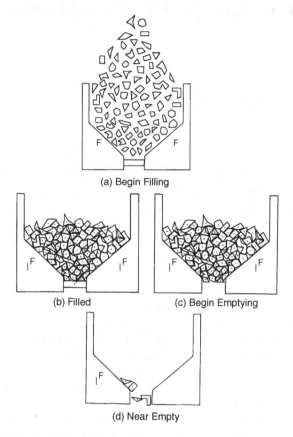

(a) Begin Filling

(b) Filled

(c) Begin Emptying

(d) Near Empty

Figure 15.1 Polygonal blocks filling and then flowing from a bin hopper in two dimensions (Cundall 1974).

15.1 DEM FORMULATION

DEM begins with preprocessing, continues with computation of block motion and deformation, and ends with post-processing. Preprocessing entails specification of initial conditions, boundary conditions, material properties, and block geometries. Computation moves blocks in time according to a finite difference scheme that requires stabilization via damping. Post-processing enables views of forces, displacements, strains, and stresses.

Computation proceeds directly from Newton's second law of motion extended to bodies of finite size. For each block, the mass center moves as a point. Thus,

$$\sum_{ext} F = ma, \quad \sum_{ext} M = I\ddot{\theta} \tag{15.1}$$

where $m, a, I, \ddot{\theta}$ are block mass, acceleration of the mass center of the block, moment of inertia about the mass center, angular acceleration of the mass center of the block, and the number of blocks. External forces F and moments M are those caused by gravity, by contact with adjacent blocks and walls and applied loads.

By definition, $a = d\dot{u}/dt$ and $\ddot{\theta} = d\dot{\theta}/dt$, where $\dot{u}$ and $\dot{\theta}$ are translational and rotational velocities of the mass center of a block. These accelerations may be approximated by central difference relationships using the velocities. Thus,

$$d\dot{u}/dt = \left[\dot{u}\left(t + \Delta t/2\right) - \dot{u}\left(t - \Delta t/2\right)\right]\Big/\Delta t, \quad d\dot{\theta}/dt = \left[\dot{\theta}\left(t + \Delta t/2\right) - \dot{\theta}\left(t - \Delta t/2\right)\right]\Big/\Delta t \quad (15.2)$$

Hence,
$$\dot{u}\left(t + \Delta t/2\right) = \dot{u}\left(t - \Delta t/2\right) + \left(\sum_{\text{ext}} F(t)/m\right)\Delta t \quad (15.3)$$
$$\dot{\theta}\left(t + \Delta t/2\right) = \dot{\theta}\left(t - \Delta t/2\right) + \left(M(t)/I\right)\Delta t$$

for the mass center of a block. Position and orientation of the ith block are estimated as

$$x_i\left(t + \Delta t/2\right) = x_i(t) + \dot{u}_i\left(t + \Delta t/2\right)\Delta t$$
$$y_i\left(t + \Delta t/2\right) = y_i(t) + \dot{v}_i\left(t + \Delta t/2\right)\Delta t \quad (15.4)$$
$$\theta_i\left(t + \Delta t/2\right) = \theta_i(t) + \dot{\theta}_i\left(t + \Delta t/2\right)\Delta t$$

The force of gravity acting on a block is simply the block weight, $W = \gamma V$, where γ is the specific weight of the block and V is the block volume. Mass of the block $M = \rho V$, where ρ is the mass density. Thus, $W/M = \gamma/\rho = g$, where g is the acceleration of gravity. To be sure, $\gamma = \rho g$.

Contact forces between blocks and between walls and blocks are considered to be proportional to displacement at a contact point. There are two types of contacts in two-dimensional DEM analysis: a corner-to-edge contact and an edge-to-edge contact, as shown in Figure 15.2. Displacements may be normal and tangential to a contact point. Thus, in incremental form at a corner-to-edge contact, $F_n = K_n u_n$ and $F_s = K_s u_s$, where K_n and K_s are constants of proportionality, that is, contact *stiffnesses*. At an edge-to-edge contact, $\sigma_n = k_n u_n$ and $\sigma_s = k_s u_s$, where k_n and k_s are *joint* stiffnesses. Associated forces are $F_n = \sigma_n L$ and $F_s = \sigma_s L$, where L is the length of the shortest edge in the edge-to-edge contact, as seen in Figure 15.2. Because blocks may be rigid, the meaning of contact displacement requires explanation. In fact, blocks may overlap during a time increment. This overlap is interpreted as displacement associated with contact force.

In an elastic system confined by rigid walls, once a block is set into motion, collision with adjacent blocks will continue indefinitely just as an elastic ball dropped on a rigid base will continue to rebound to the original height indefinitely. Of course, this action does not happen in reality. Dissipation of energy occurs (as heat transferred to surroundings), the motion is *damped*, and the ball rapidly comes to rest. Damping in DEM is necessary for realism and numerical stability. Two types of viscous damping may be used in DEM, either alone or together. The resulting damping force F_d opposes the motion.

The equation of translational motion is then

$$F = m\left(du^2/dt^2\right) + mg - F_d \quad (15.5)$$

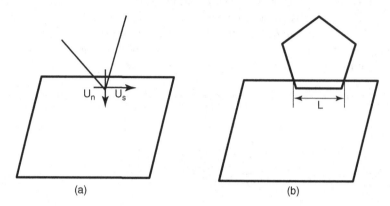

Figure 15.2 Contacts in two dimensions: (A) corner to edge and (B) edge to edge.

Substitution of (15.5) into (15.3) allows for computation of velocities and new positions of blocks as in (15.4).

The finite difference computation of velocities, displacements, and rotations outlined previously appears deceptively simple. However, the critical task of determining contacts is not so simple. Indeed, the task of contact detection consumes most of the computation time in DEM analysis despite clever storage schemes for linking block data geometries and nearest neighbors. But with the advance of computer size and speed with reduction in hardware costs, block and particle DEM analyses have advanced considerably and have found numerous practical applications.

An early example of a practical application of *block* DEM analysis is given by Ryu (1989) who adapted a program by Cundall et al. (1978) to cast blasting in surface mining of coal. The objective of cast blasting is to move overburden away from the underlying coal and thus to avoid loading and transporting the overburden by shovel and truck operation. Figures 15.3 and 15.4 are example plots of results.

The number of blocks in this example is about 145. Run times varied with the computer used from almost 2 h on a Gould to about 15 min on a CRAY XMP/48. Much faster run times would be expected in consideration of the advances in computer speed.

A practical example of *particle* DEM is given by Acharya (1991). This example concerns ball milling in the metallurgical industry and is of importance because of energy consumption in reducing ore particle size. Figure 15.5 illustrates results in graphical form. The program, 3DBALL, was developed from the open-source program TRUBAL (Cundall and Strack 1979). An early application to comminution in ball mills is described by Mishra et al. (1990). Subsequent development by colleagues in metallurgical engineering advanced the technology to the point of commercial software with numerous applications since the pioneering effort by Acharya (1991) in ball milling.

The state of the art of particle DEM applications to materials handling problems involving millions of particles may be found at the website for EDEM. Of course, there are other software vendors offering DEM solutions to industrial problems and to problems of public safety involving rockfalls along transportation routes.

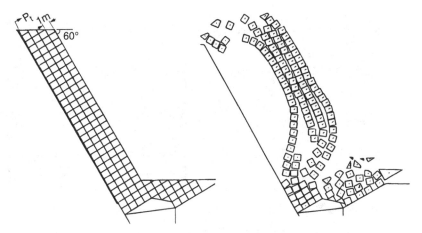

Figure 15.3 Inclined blast hole model. Bench height is 20 m and dips at 60°. Expansion of the gas bubble in the blast hole propels the blocks: (A) before detonation and B) 69.6 ms after detonation. Joints are spaced 1 m and dip at 30° and 60°. Blocks are thus 1 m square in this two-dimensional DEM analysis (Ryu 1989).

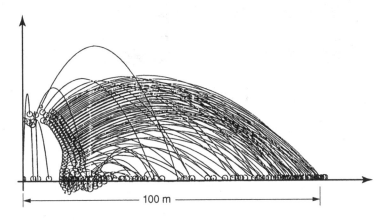

Figure 15.4 Block trajectories. Maximum throw is approximately 102 m (Ryu 1989).

About 500 balls were necessary to fill the mill to 50% capacity, (C) and (D) in Figure 15.5. Run time for the ball simulation on a SPARC 1+ station was about 18 h.

15.2 DDA FORMULATION

Discontinuous deformation analysis (DDA) begins as a technique for determining displacements that best fit measured displacements of a blocky rock system (Shi and Goodman 1984, 1985). Displacements included translation and rotation shared by all points in a block and displacements in a neighborhood of a point associated with volume and shape changes. Blocks in a system are allowed to separate, but not overlap.

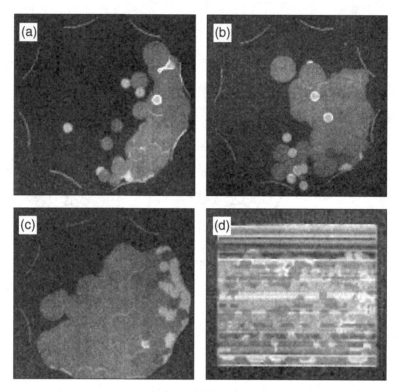

Figure 15.5 Snapshots of ball mill simulation by DEM (Acharya 1991). (A) Cascading at 60% of critical speed at time = 0.5631 s. (At time = 0.5928 s. (C) Front view of the ball mill at 1.2447 s. (D) Side view of the ball mill at 1.2447 s (Acharya 1991). There are four ball sizes in the snapshot that are easily visible in color compared with grayscale images.

In two dimensions, the displacements u and v in the x- and y-directions are

$$\begin{Bmatrix} u \\ v \end{Bmatrix} = \begin{bmatrix} 1 & 0 & -(y-y_o) & (x-x_o) & 0 & (y-y_o) \\ 0 & 1 & -(x-x_o) & 0 & (y-y_o) & (x-x_o) \end{bmatrix} \begin{Bmatrix} u_o \\ v_o \\ \omega \\ \varepsilon_{xx} \\ \varepsilon_{yy} \\ \gamma_{xy} \end{Bmatrix} \tag{15.6}$$

where u_o, v_o, ω, ε_{xx}, ε_{yy}, γ_{xy} are displacements of the mass center of a block, block rotation about the mass center, normal strains, and shear strain. In compact form, (15.6) is

$$\begin{Bmatrix} u \\ v \end{Bmatrix}_{2\times 1} = [k]_{2\times 6} \{d\}_{6\times 1} \tag{15.7}$$

The squared deviation of measurements m_1 and m_2 of displacements is

$$\Phi_a = (m_1 - u)^2 + (m_2 - v)^2 \tag{15.8}$$

Consider a point $p_1(x, y)$ in block i and $p_2(x, y)$ in block j, and let m be the extension of a line between the two points. Let p^i and p^j be projections of displacement on the line between the two considered points. The squared deviation of extension between the two points is then

$$X_b = \left[m + \left(p^j - p^i \right) \right]^2 \tag{15.9}$$

Also consider a block corner $c(x, y)$ at the end of a block edge shared by blocks i and j and displacements normal to the edge of each block, e_i, e_j ("extensions"). If $e_i > -e_j$, then the edges tend to separate at $c(x, y)$. If $e_i < e_j$, the edges tend to close at $c(x, y)$. Preventing excess closure is done with a penalty function. Thus,

$$\Psi_c = (e_1 + e_2)^2 P \tag{15.10}$$

where $P = 0$ if $(e_i + e_j) = 0$ and $P =$ a large number if $(e_i + e_j) \geq 0$.

Considering all blocks, the least square function to be minimized is

$$\sum_a \Phi_a + \sum_b X_b + \sum_c \Psi_c \tag{15.11}$$

for all measurement points a, b, c. The result of minimization generates a system

$$\{F\} = [K]\{D\} \tag{15.12}$$

that consists of $6n$ equations in $6n$ unknowns (Shi and Goodman 1984). Solution of the system (15.12) requires three points per block and is estimated to require about five times the run time a comparable finite element analysis would require ($3n$ equations in $3n$ unknowns). To be sure, inversion of (15.11) gives a least square best fit to the measured block system "displacements" in terms of the right-hand side of (15.6) at each point.

For engineering design, a forward analysis that anticipates motion in a blocky rock system is needed rather than a back-analysis of fitting theory to observations. Details of both are given by Shi (1988) with examples and sketches illustrating the geometry of the many features that enter the analysis forward and backward. Figure 15.6 is an illustrative example of block motion about a tunnel under symmetric stress.

Considerable advancement in DDA technology has followed the original work by Shi and Goodman (1984), aided by the enormous advancements in digital technology. International conferences on DDA contain a wealth of information concerning new technology and applications.

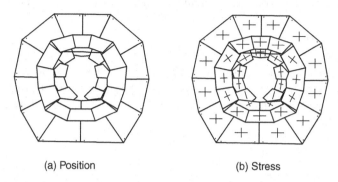

(a) Position   (b) Stress

Figure 15.6 A DDA example analysis of block motion about a tunnel (Shi 1988).

REFERENCES

Acharya, A. (1991) "A Discrete Element Approach to Ball Mill Mechanics". M.S. Thesis, University of Utah, pp 103.

Cundall, P.A. (1971) A Computer Model for Simulating Progressive Large Scale Movements in Blocky Rock Systems. *Proceedings of the International Symposium on Rock Fracture*, Nancy (ISRM), Paper 2–8.

Cundall, P.A. (1974) *Rational Design of Tunnel Supports: A Computer Model for Rock Mass Behavior Using Interactive Graphics for the Input and Output of Geometrical Data.* Final Report: Department of the Army Contract No. DACW 45-74-C-0066, Geology, Soils, and Materials Branch, Missouri River Division, Corps of Engineers, Omaha, NE.

Cundall, P.A., J. Marti, P. Beresford, N. Last and M. Asgian (1978) *Computer Modelling of Jointed Rock Masses.* Final Report Dames and Moore Contract No. DACA39-77-0004, Defense Nuclear Agency, U.S. Army Engineer Waterways Experiment Station Weapons Effects laboratory, Vicksburg, MS.

Cundall, P.A. and L. Strack (1978) *The Distinct Element Method as a Tool for Research in Granular Media (Part 1).* Report to the National Science Foundation Concerning NSF Grant Eng 76-20711, pp 232.

Cundall, P.A. and L. Strack (1979) *The Distinct Element Method as a Tool for Research in Granular Media (Part 2).* Report to the National Science Foundation Concerning NSF Grant Eng 76-20711, pp 203.

Cundall, P.A. and R.D. Hart (1985) *Development of Generalized 2-D and 3-D Distinct Element Programs for Modeling Jointed Rock.* Final Report. Contract No. DACA 39-82-C-0015. U.S. Army Corp of Engineers, Vicksburg Waterways Experiment Station.

Jing, L. and O. Stephansson. (2007) *Fundamentals of Discrete Element Methods for Rock Engineering: Theory and Applications.* Elsevier Science, Amsterdam, pp 563.

Mishra, B.K., R.K. Rajamani and W.G. Pariseau (1990) Simulation of Ball Charge Motion in Ball Mills. *Society of Mining, Metallurgical and Petroleum Engineers Annual Meeting*, Salt Lake City, Utah. PrePrint 90-137.

O'Sullivan, C. (2011) *Particulate Discrete Element Modelling: A Geomechanics Perspective.* Spon Press, London and New York, pp 561.

Ryu, C.-H. (1989) "Numerical Modeling of Fragmentation in Jointed Geologic Media." Ph.D. Dissertation, University of Utah, pp 201.

Shi, G.-H. and R.E. Goodman (1984) Discontinuous Deformation Analysis. *Proceedings of the 25th Symposium on Rock Mechanics.* SME/AIME, New York. pp 269–277.

Shi, G.-H. and R.E. Goodman (1985) "Two-dimensional Discontinuous Deformation Analysis." *Int. J. Num. Anal. Methods Geomech.* Vol. 9, No. 6, pp 541–556.

Shi, G.-H. (1988) "Discontinuous Deformation Analysis: A New Numerical Model for the Statics and Dynamics of Block Systems." Ph.D. Dissertation, University of California, Berkeley, pp 379.

Chapter 16

Conclusion

These "Notes" began with an intuitive approach to the finite element method (FEM) via interpolation over a triangle. This approach is contrary to the approach found in many reference books on FEM where an energy principle in integral form is subject to minimization. However, the intention here is to ease the learning process rather than to impress the reader. Focus is on solid mechanics, especially the mechanics of geological media, rocks, and soils. Consideration of porous media is essential and introduces fluid flow and the key concept of effective stress. For convenience, a review of fundamental concepts is provided in an appendix.

The triangle is two-dimensional, of course, and avoids over-simplification found in one-dimensional elements and over-complication using a three-dimensional element. However, the triangle, while useful for learning, is not a numerically efficient finite element. A quadrilateral element is a much more useful element and allows for introduction of isoparametric elements and shape functions that turn out to be equivalent to the interpolation functions introduced earlier.

Use of the divergence theorem as a principle of virtual work leads to element equilibrium of a linearly elastic body in the form of element stiffness equations. Numerical forms for element strains and stresses follow. Global equilibrium considering all elements in a mesh follows logically from addition of element forces.

A quadrilateral formed by four constant-strain triangles provides an alternative to a conventional 4-node quadrilateral. Elimination of the central node common to the four individual triangles introduces static condensation and equation solving. This quadrilateral also proves to be useful in analyses of elastic–plastic solids, including saturated, porous, fractured solids.

Equation solving by elimination and iteration is then discussed. Boundary conditions are introduced in several ways. Advantages of Gauss elimination and Gauss–Seidel iteration with over-relaxation are described within the concept of bandwidth and computer capacity. Mention is made of the popular conjugate gradient iterative solver.

Material nonlinearity arises when the elastic limit is reached and loading continues. Equation solving then involves an incremental approach using a tangent stiffness scheme or a modified Newton–Raphson scheme.

All real processes take time, although time does not enter explicitly into elastic and elastic–plastic analyses as such. However, consideration of time is unavoidable in poroelastic and poroelastic–plastic analyses. This consideration involves an explicit time-stepping scheme, perhaps a forward, backward, or Crank–Nicholson scheme.

DOI: 10.1201/9781003166283-16

In case of nonlinearity, coincidence of load steps and time steps is a considerable numerical convenience, although not a requirement.

Interpolation of pressure and temperature is carried forward with much the same formulation as for solid displacement. Consideration of seepage according Darcy's law is discussed in cases of incompressible flow through rigid (undeforming) porous solids and compressible flow through deformable solids. Element stiffness is then obtained through the divergence theorem in the form of a principle of virtual power (work rate). Rates and transient analysis imply explicit time dependency, of course.

Compressible fluid flow in deforming porous solids entails hydro-mechanical coupling, where the concept of effective stress is essential to understanding. Elucidation of the process leads to an extension of Hooke's law for elastic solids to Biot's law for poroelastic solids. Limiting the range of elastic deformation by strength then allows for poroelastic–plastic analysis.

Brief discussions of the boundary element method (BEM) and the distinct element method (DEM) as well as discontinuous deformation analysis (DDA) are given. Mention of advantages and disadvantages relative to FEM is made.

These "Notes" are intended for class and self-study. Accordingly, problems for weekly assignments over a 15-week term of study (semester) are provided in an appendix. These problems involve questions about numerical concepts and later numerical analyses using software available from outside sources. Solutions to these problems are given in another appendix.

There are a number of companies and commercial websites that provide software for FEM, BEM, DEM, and DDA. Some provide student versions at no cost or at an educational discount. There are also programs to be found as freeware on the Internet where caution is in order.

In conclusion, while one may generate an intellectual understanding, say, of FEM, by reading and running existing computer programs, the most informative approach in the author's experience and as remarked upon by Akin (1994) is a hands-on approach, writing and running one's own computer code as suggested in the "Notes".

REFERENCE

Akin, J.E. (1994) *Finite Elements for Analysis and Design*. Academic Press, New York, pp 548.

Appendix A

Review of fundamental concepts

In a broad sense, what one attempts to do in rock engineering is to anticipate the motion of a proposed structure under a set of given conditions. Thus, the main design objective is to calculate displacements and, as a practical matter, to see whether the displacements are acceptable. Very often, restrictions on displacements are implied rather than stated outright. This situation is almost always the case in *elastic* design where the displacements are restricted only to the extent that they remain within the range of elastic behavior. A tacit assumption is that in the absence of inelastic behavior, large displacements accompanying failure are precluded. Under these circumstances, design is essentially an analysis of safety and stability. The design objective is then to calculate a factor of safety appropriate for the problem at hand.

In rock mechanics, the "structure" of interest is simply the rock mass adjacent to a proposed excavation. The proposed excavation may be started at the surface, or it may be a deepening of an existing surface excavation, a start of a new underground excavation, or an enlargement of an existing underground excavation. In any case, the excavation plan, if actually carried out, would cause changes in the forces acting in the neighborhood of the excavation and would therefore cause deformation of the adjacent rock mass. Of course, the rock mass may also move in response to other forces such as those associated with earthquakes, equipment vibration, blasting, construction of a nearby waste dump, filling or draining of an adjacent water reservoir, and temperature changes.

An appropriate factor of safety in design depends on the situation and in any event is an empirical index to "safety" or "stability". Safety and stability are not strictly synonymous in mechanics of solids, but are often used interchangeably in problem discussions. If forces are of primary concern, then a ratio of forces resisting motion to forces that tend to drive the motion is an appropriate safety factor. If rotation is of primary concern, then a ratio of resisting to driving moments would be an appropriate safety factor. When yielding at a point is of interest, then a ratio of "strength" to "stress" defines a useful safety factor when measures of "strength" and "stress" are well defined.

Regardless of the specific identity of the forces acting, the associated motion must always be consistent with basic physical laws such as the conservation of mass and the balance of linear momentum. In this respect, rock is no different than other materials. Any motion must also be consistent with the purely geometrical aspects of translation, rotation, change in shape, and change in volume, that is, with kinematics. However, physical laws such as Newton's second law of motion (balance of linear momentum) and kinematics are generally not sufficient for the description of the motion of a deformable body. The number of unknowns generally exceeds the number of equations.

This mathematical indeterminacy may be removed by adding to the system as many additional equations as needed without introducing additional unknowns. The general nature of such equations becomes evident following an examination of the internal mechanical reaction of a material body to the externally applied forces. The concepts of stress and strain arise in such an inquiry, and the additional equations needed to complete the system are equations that relate stress to strain. Stress–strain relationships represent a specific statement concerning the nature or "constitution" of a material and are members of a general class of equations referred to as constitutive equations. Constitutive equations express material laws. Whereas physical laws and kinematics are common to all materials, constitutive equations serve to distinguish categories of material behavior. Hooke's law, for example, characterizes materials that respond elastically to load. A system of equations that describes the motion of a deformable body necessarily includes all three types of equations: physical laws, kinematics, and material laws.

In reality, a system describing the motion of a material body is only an approximation. Mathematical complexities often dictate additional simplification and idealization. Questions naturally arise as to what simplifications should be made and, once made, how well the idealized representation corresponds to reality. Questions of this type relate more to the art than to science of engineering design and have no final answers. Experience can, of course, be a great aid in this regard, provided such experience is informed by a clear understanding of the fundamental concepts.

I   PHYSICAL LAWS

Physical laws are common to the mechanics of all materials. Regardless of the constitution of the material, mass must always be conserved and the balances of linear and angular momentum must be maintained. To these laws, one may add the balances of energy and entropy, the first and second laws of thermodynamics. In thermodynamics, an important distinction is made between control volume and control mass analyses. The first allows for flow to and from a region of interest, a volume in space (which may deform); the latter is a fixed, specified mass, that is, a material body (which may also deform). Physical laws apply to material particles and material bodies. However, the conservation of mass and the balances of momentum hold independently of the balances of energy and entropy. A purely mechanical analysis is therefore feasible without explicit treatment of thermodynamic questions.

1.1   Conservation of mass

Let dm be a small mass element in a material body B that occupies volume V, as shown in Figure 1.1, then the mass M of B is

$$M = \int_V dm \tag{1}$$

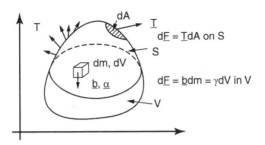

Figure 1.1 A material body under external loads.

Conservation of mass of the considered body requires M to remain constant, so

$$\frac{dM}{dt} = \frac{d}{dt}\int_V dm = \int_V \frac{d(dm)}{dt} = 0 \tag{2}$$

The assumption that mass is distributed, that is, continuous to an infinitesimal scale, allows for a definition of mass density; thus, $\rho = dm/dV$. Hence,

$$M = \int_V \rho \, dV \tag{3}$$

Now, however,

$$\frac{dM}{dt} = \frac{d}{dt}\int_V \rho \, dV = \int_V \frac{\partial \rho}{\partial t} \, dV + \int_S \rho v_n \, dS \tag{4}$$

where v_n is the velocity at a point on the surface S of V in a direction outward and normal to S. The first integral accounts for change in mass at a fixed V, while the second term accounts for S moving as the body deforms and V changes with time. If V were a fixed region in space, then the surface integral would vanish.

1.2   Balance of linear momentum

The correct form of Newton's second law of motion for a particle is

$$F = \dot{P} \tag{5}$$

where the superior dot means time derivative, and F and dP/dt are vectors of resultant of *external* forces and time rate of change of linear momentum, respectively.

The linear momentum of an element of mass of the body is the product of velocity times mass, vdm; the total momentum of the body is just

$$P = \int_V v \, dm \tag{6}$$

Velocity is the time rate of change of position, of course. If r is a position vector, shown in Figure 1.2, then $dr/dt = v$; acceleration is $\dot{v} = a$ and

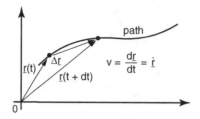

Figure 1.2 Particle position and velocity.

$$\dot{P} = \int_V a\, dm = \int_V \rho a\, dV \tag{7}$$

when the mass balance is also satisfied.

Two types of external forces contribute to the left-hand side of (5): contact forces T distributed on S (surface tractions) and body forces distributed in V, that for practical purposes are just gravity forces as shown in Figure 1.1. The resultant of these distributed forces is

$$F = \int_S T\, dS + \int_V \gamma\, dV \tag{8}$$

where γ is weight per unit volume (specific weight) of the material; $\gamma = \rho g$, in which g is the acceleration of gravity. The balance of linear momentum, in essence, Newton's second law of motion, is then

$$\int_S T\, dS + \int_V \gamma\, dV = \int_V \rho a\, dV \tag{9}$$

If one is given the applied loads, then the left-hand side of (9) is known. However, to determine the motion of the body, one needs to find the acceleration that is under the integral sign on the right-hand side of (9). The acceleration varies from point to point. In essence, there are an infinite number of unknowns, but only two equations (conservation of mass and balance of linear momentum). If the acceleration field were known, then a time integration would give the velocity field. A second time integration would give the desired displacements.

The mass center of a body is defined by

$$Ms = \int_V r\, dm \tag{10}$$

where s is the position vector of the mass center as shown in Figure 1.3. Differentiation of (10) twice with respect to time gives the interesting result

$$M\ddot{s} = \int_V a\, dm = \int_V \rho a\, dV = \dot{P} = F \tag{11}$$

that shows that the mass center moves as if it were a particle accelerating according to Newton's second law. Thus, even though one cannot determine the acceleration everywhere in the body of interest at this juncture, there is a possibility of at least following

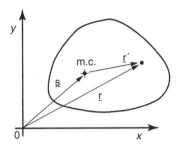

Figure 1.3 Center of mass and position relative to the mass center.

the motion of the center of mass of the body. This fact remains true even if the body disintegrates.

Example 1.1

Consider a mass M of rock in a landslide or avalanche, and suppose that the external forces are (i) the weight W of the slide mass and (ii) the contact force R acting between the slide mass and the parent rock mass from which the slide mass has become detached. The contact force R may be frictional, viscous, and displacement dependent, that is, $R = R_o + R_1 \, ds/dt - R_2 s$. R increases with speed, but decreases with displacement of the mass center and resists the downhill component of weight $D = D_o$. According to (11),

$$F = D - R = M\ddot{s}$$

Hence,

$$M\ddot{s} = R_1\dot{s} - R_2 s = D_o - R_o$$

describes the motion of the slide mass center. The coefficients in this equation may depend on time and position (except M). Over a short period of time, however, a reasonable assumption is that they are constant in which case the form of the solution is known. Reasonable initial conditions are that the slide mass is at rest or moving at a constant speed (steady creep).

The logic followed here illustrates in a very compact form how one proceeds from physical laws (conservation of mass and Newton's second law) through problem simplification (look at mass center motion only; disregard deformation, disintegration, and motion of individual elements) and material idealization (assumptions concerning resistance) to a mathematically tractable representation of the original problem (landslide dynamics). Of course, simplification is a relative notion. Here, simplification means one has progressed from an essentially hopeless situation to a situation where useful information may be extracted. In this example, useful information might refer to estimation of slide mass travel in conjunction with zoning regulation for geologic hazard. The solution effort required may still be considerable. However, there may also be unexpected benefits. In the present example, "triggering" of catastrophic landslides under load level fluctuations that were formerly safe becomes understandable in relatively simple physical terms.

1.3   Balance of angular momentum

The balance law for angular momentum states that the resultant external moment (torque) L acting on a body is equal to the time rate of change of angular momentum dH/dt. Thus,

$$L = \dot{H} \tag{12}$$

expresses the balance of angular momentum. The resultant moment is the sum of the moments exerted on the body by the external forces distributed over the surface and throughout the volume of the body, while the time rate of change of angular momentum is just the moment of linear momentum. Thus,

$$\int_S (r \times T)\,dS + \int_V (r \times \gamma)\,dV = \int_V (r \times a)\,dm \tag{13}$$

Here, r is the position of dS on S where T acts, the position of dV and dm in V and $\times$ means vector product (cross product).

1.4   Balance of energy

The balance of energy as a physical law has a degree of arbitrariness not found in the conservation of mass and the momentum balances. Like all balance laws, the balance of energy for a material body can be stated as (increase) = (supply); that is, the net increase in the energy of a body, say, per unit of time, must equal the net amount of energy supplied to the body (per unit time) from external sources. If dE/dt is the time rate of change of energy in the body, dW/dt is the time rate of doing work on the body and dQ/dt is the rate of heat addition to the body; the balance law for energy in rate or power form is

$$\dot{E} = \dot{W} + \dot{Q} \tag{14}$$

Kinetic energy K is included in the total energy E. Other types of energy such as energy of deformation, surface energy, and chemical energy may also be identified individually and separated from E. The remaining energy U is the "internal" energy. For example, after separating K from E, we have $U = \dot{W} + \dot{Q} - \dot{K}$. If the heat transfer rate is negligible, then $\dot{W} = \dot{U} + \dot{K}$ shows that in a purely mechanical energy balance, the work done by the external forces increases the internal energy and accelerates the body. In this case, the internal energy rate is just the energy associated with deforming the body, so $\dot{W}_{ext} = \dot{W}_{def} + \dot{K}$, where for emphasis the subscripts ext = external and def = deformation have been added. If there is frictional slip associated with relative motion between parts of a body, then

$$\dot{W}_{ext} = \dot{W}_{def} + \dot{W}_{slip} + \dot{K} \tag{15}$$

The external forces do work on a body at the rate

$$\dot{W}_{ext} = \int_S (T \cdot v)\,dS + \int_V (\gamma \cdot v)\,dV \tag{16}$$

where (·) means scalar (dot or inner) product. By definition, the kinetic energy is

$$K = \int_V (1/2)(v \cdot v)\,dm \tag{17}$$

that has the rate $\dot{K} = \int_V (v \cdot a)\,dm$.

The kinetic energy of a body can be separated into two parts: one part associated with the mass center motion and the other part associated with motion relative to the mass center. From (17),

$$\dot{K} = \int_V (1/2)(\dot{s} + \dot{r}')(\dot{s} + \dot{r}')\,dm$$

where s is the position of the mass center and r' is the position relative to the mass center, that is, $r = s + r'$. After carrying out the integration, one obtains

$$\dot{K} = (1/2)m(\dot{s})^2 + \int_V (1/2)(v' \cdot v')\,dV$$

where v' is velocity relative to the mass center located at s. If a body is considered to be rigid, then the motion is a translation of the mass center and a rotation about the mass center. In the general case, the motion includes deformation as points of the body move relative to each other. Detailed consideration of the purely geometrical aspects of a motion is the subject of kinematics.

2   ANALYSIS OF STRESS

Stress is the internal mechanical reaction of a deformable body to externally applied forces. The natural starting point for an investigation of stress is at the surface of a body where the external forces lie to the outside and are known. The internal forces lie to the inside and are to be determined from equilibrium requirements. The analysis is thus based on established physical law. In choosing a point on the surface of a body, one is selecting a special point that may not be representative of interior points. A check must therefore be made to see whether the characterization of the internal forces remains valid. Focus on a surface point is intuitively logical, while the investigation of an interior point is rigorous and leads to definition of stress vector, state of stress, and stresses. All agree with the results obtained from the study of the generic surface point. A two-dimensional analysis contains all the essential features of a three-dimensional view and affords an economy of thought and a welcome reduction in the number of terms appearing in the analysis.

2.1 Surface tractions and stresses

Figure 2.1 shows the external forces acting on a small surface element at a typical point P on the surface of the considered body. A small triangular element is isolated for analysis in Figure 2.2. The external force is a surface traction T. Orientation of the surface element is indicated by the unit outward normal vector $\mathbf{n}$. The association of the surface traction and orientation is emphasized with a superscript n. All forces shown in the free body diagram in Figure 2.2 are distributed loads. The total force acting at P is $dF = TdA$; in words, (force) = (force per unit area) (area); dF and T have

Thus,

$$\sum F_x = 0 = T_x^n dA - T_x^x dA_x - T_x^y dA_y$$
$$\sum F_y = 0 = T_y^n dA - T_y^x dA_x - T_y^y dA_y \tag{18}$$

The leading terms in (18) can be obtained by first multiplying the traction by the associated area to obtain the total force and then resolving into components, that is,

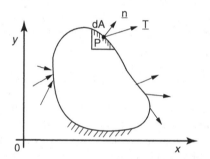

Figure 2.1 Surface forces or tractions, details in the neighborhood of a surface point with double-subscript notation.

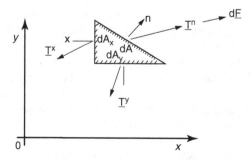

Figure 2.2 Surface forces or tractions, details in the neighborhood of a surface point with double-subscript notation.

$$dF_x = dF\cos(\beta) = TdA\cos(\beta) = T_x^n dA$$
$$dF_y = dF\sin(\beta) = TdA\sin(\beta) = T_y^n dA$$

(19)

Other terms in (18) can be obtained in the same way (traction) (area) (cos or sin). From the geometry of Figure 2.3,

$$dA_x = dA\cos(\alpha), \; dA_y = dA\sin(\alpha)$$

(20)

which can be used to eliminate dA from (18). Thus,

$$T_x^n = T_x^x \cos(\alpha) + T_x^y \sin(\alpha)$$
$$T_y^n = T_y^x \cos(\alpha) + T_y^y \sin(\alpha)$$

Stresses are now defined as the components of tractions acting on surfaces parallel to the coordinate planes. Thus, with a change to a double-subscript notation, the stresses are

$$\sigma_{xx} = T_x^x, \; \tau_{xy} = T_y^x, \; \tau_{yx} = T_x^y, \; \sigma_{yy} T_y^y$$

(21)

where the first subscript indicates the normal direction and the second subscript indicates the action direction as shown in Figure 2.4. The symbol σ ("sigma") is used for stresses acting perpendicular to the considered area; τ is used for stresses acting parallel to the considered area. The former are *normal* stresses; the latter are *shear* stresses.

The direction of the normal in (18) is implied by the angle *a*, so the superscript *n* can be omitted from the terms on the left. Equations (18) can be refined somewhat more by noting that the normal direction can be specified by direction cosines. With $\cos(\alpha) = \cos(n,x)$ and $\sin(\alpha) = \cos(n,y)$, one has

$$T_x = \sigma_{xx}\cos(n,x) + \tau_{xy}\cos(n,y)$$
$$T_y = \tau_{yx}\cos(n,x) + \sigma_{yy}\cos(n,y)$$

(22)

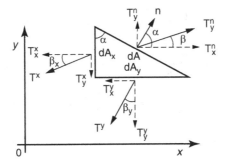

Figure 2.3 Surface forces or tractions, details in the neighborhood of a surface point with double-subscript notation.

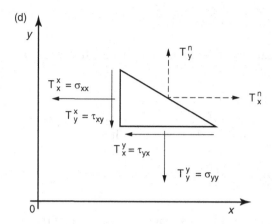

Figure 2.4 Surface forces or tractions, details in the neighborhood of a surface point with double-subscript notation.

A three-dimensional analysis based on a tetrahedron, as shown in Figure 2.5, leads to

$$T_x = \sigma_{xx} \cos(n,x) + \tau_{yx} \cos(n,y) + \tau_{zx} \cos(n,z)$$
$$T_y = \tau_{xy} \cos(n,x) + \sigma_{yy} \cos(n,y) + \tau_{zy} \cos(n,z)$$
$$T_x = \tau_{xz} \cos(n,x) + \tau_{yz} \cos(n,y) + \sigma_{zz} \cos(n,z)$$

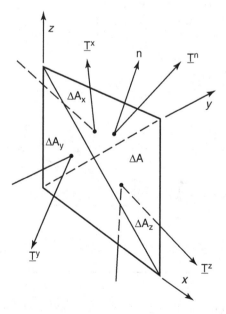

Figure 2.5 Tractions acting on a tetrahedral neighborhood of a point.

that differ from the two-dimensional relations between tractions and stresses only by the z-direction terms. The direction cosines are just the components of an outward normal vector to the considered area; this vector has unit length. If this vector is $\boldsymbol{n}$ with components (n_x, n_y, n_z), then

$$
\begin{aligned}
T_x &= \sigma_{xx}n_x + \tau_{yx}n + \tau_{zx}n_z \\
T_y &= \tau_{xy}n_x + \sigma_{yy}n_y + \tau_{zy}n \\
T_x &= \tau_{xz}n + \tau_{yz}n_y + \sigma_{zz}n_z
\end{aligned}
\tag{23}
$$

which differ from the two-dimensional analysis and consider the neighborhood of a surface point and require equilibrium there.

2.2　Stress vector and state of stress

The previous analysis considered the neighborhood of a surface point and required equilibrium there. Sectioning of a body now creates a surface that was interior to the body. Redoing the analysis for a point on this surface amounts to requiring equilibrium at an interior point. The formal result is (23) with the distinction that previously the components on the left-hand side were components of a (surface) traction, but now are components of a *stress vector*. The process is illustrated in Figure 2.6a.

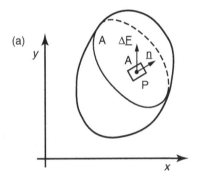

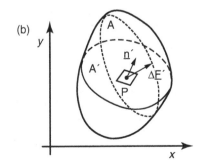

Figure 2.6 Two surface elements at the same point inside a material body.

The distributed force at the considered point, the stress vector, may also be obtained by a limiting process

$$\lim_{\Delta A \to 0} \left(\frac{\Delta F}{\Delta A} \right) = T \tag{24}$$

The stress vector depends on the considered point P. At some other point P′, the stress vector will have a different value. However, because the point is in the interior of the body, one can section the body at the same point, but at a different orientation, as shown in Figure 2.6b. The stress vector would be different even though the location has not changed. Hence, the stress vector depends on position and orientation, that is, $T = T(P, \boldsymbol{n})$. The set of stress vectors at P defines the *state of stress*. This definition is not very practical because there is an infinite number of stress vectors at an interior point, one for every orientation n.

2.3   Cauchy stress formula and sign convention

When Equations (23) are derived for an interior point, the result is the famous Cauchy stress formula. This formula reduces the specification of the state of stress to just nine numbers, the stresses previously defined, because a stress vector at any given orientation can be computed by Formula (23). The magnitude of the stress vector shown in Figure 2.7a is

$$T = \sqrt{T_x^2 + T_y^2 + T_z^2} \tag{25}$$

with direction cosines

$$\cos(T, x) = \frac{T_x}{T}, \cos(T, y) = \frac{T_y}{T}, \cos(T, z) = \frac{T_z}{T} \tag{26}$$

that are generally different for the normal direction, of course. In fact, the angle between the stress vector and the normal to the considered area can be obtained in the usual way that the angle between two lines in space is obtained. If e is this angle, then

$$\cos(\theta) = \cos(T, x)\cos(n, x) + \cos(T, y)\cos(n, y) + \cos(T, z)\cos(n, z)$$

The component of the stress vector perpendicular to the considered area may be obtained from the dot product $T \cdot n$. Thus,

$$N = T_x n_x + T_y n_y + T_z n_z = T\cos(\theta) = \sigma \tag{27}$$

is a normal stress relative to a coordinate direction parallel to $\boldsymbol{n}$ as shown in Figure 2.7b. The component parallel to the considered area is just

$$S = \sqrt{T^2 - N^2} = T\sin(\theta) = \tau \tag{28}$$

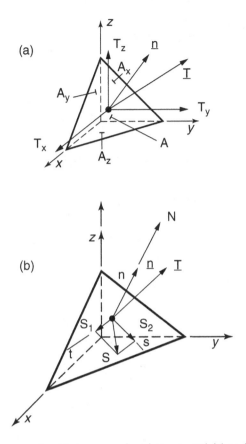

Figure 2.7 Stress vector resolved into normal and tangential (shear) components.

r stress relative to a coordinate direction parallel to the considered surface and in the plane of T and n. Alternatively, the shear component S may be resolved into two shears S_1 and S_2 parallel to the considered surface (but not in the plane of T and n), as shown in Figure 2.7b. If one labels the normal direction and the two shear directions as n, s, and t that, by choice, are mutually orthogonal, then $\sigma = \sigma_{nn}$, $S_1 = \tau_{ns}$, $S_2 = \tau_{nt}$. Thus, on any surface there are generally two shear stresses and a normal stress.

The sign convention that makes tension positive is shown in Figure 2.8a. The shear stresses are also positive on the faces of the cube in Figure 2.8a. The convention that makes all stresses positive in Figure 2.8a comes from the sense of the subscripts. If the two subscripts have the same sense, both positive or both negative, then the considered stress is positive. If compression is considered positive, the sign is that the considered stress is positive when the two subscripts are opposite in sense. This sign convention is shown in Figure 2.8b.

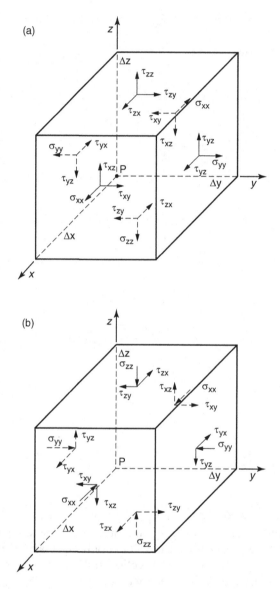

Figure 2.8 Sign convention for stress: (a) tension positive and (b) compression positive.

Example 2.1

An example calculation using some of the previous results illustrates potential value. Suppose the state of stress at some point below the surface of the earth is given such that the vertical stress is the product of the average specific weight of rock multiplied by depth (a very common estimate). The horizontal stresses are expected to be some fraction of the vertical stress when only weight of rock is considered. To be definite, take the z-direction to be vertical and positive upward from the surface, the positive x-direction to be east, and the positive y-direction to be north, and call this system of reference *compass coordinates*. With tension positive, the vertical stress relative to this system is just $\sigma_{zz} = \gamma z$, where γ is the specific weight of rock, say, 144 pounds per cubic foot. At a depth of 1,500 ft, $\sigma_{zz} = -1,500$ (z is negative down, and compression is negative). Assume that the horizontal stresses are equal and $\sigma_{xx} = \sigma_{yy} = -300$ pound per square inch and no shear stresses are present. These are reasonable assumptions for a rock mass loaded by gravity only. Suppose a fault that bears north 30° and dips 60° southeast passes through the considered point. If the absolute value of the shear stress $|\tau|$ is less than the shear strength τ_f of the fault, then slip does not impend. The absolute value is used because the sign only indicates direction. Does slip impend? Or what is the factor of safety with respect to frictional slip at the point of interest? Is $FS = \tau_f / |\tau| > 1$?

One may suppose that fault resistance to slip is purely frictional so that $\tau_f = \mu\sigma$, where μ is the coefficient of fault friction and σ is the magnitude of the normal (compressive) stress on the fault at the considered point. Normal and shear stresses on the considered fault are illustrated in Figure 2.9. Solution to the problem requires calculation of σ, τ, and τ_f. The coefficient of friction is estimated to be 0.7, which corresponds to an angle of friction of 35°, $\mu = \tan(\phi)$. Calculations can proceed, provided one can find the direction cosines n_x, n_y, and n_z of the fault plane normal.

Finding the direction cosines of an outward unit vector normal to a fault or joint is a problem frequently encountered in rock mechanics, so a digression is worthwhile at this juncture. A somewhat more general problem is that of computing a rotation of reference axes from compass coordinates (x, y, z) to fault or *joint plane coordinates* (a, b, c). These latter coordinate directions are down dip, parallel to strike, and normal to the considered plane. Figure 2.10 shows the geometry of the problem. Dip angle o is positive down from the horizontal; the dip direction a is positive clockwise from north. The dip direction is an azimuthal angle measured in a horizontal plane. Strike of the considered plane is a-90o. Expressions for the direction cosines between compass and joint plane coordinate directions can be determined from the geometry of Figure 2.10 and are given in Table 2.1.

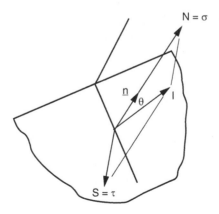

Figure 2.9 Normal and shear stresses on a fault plane.

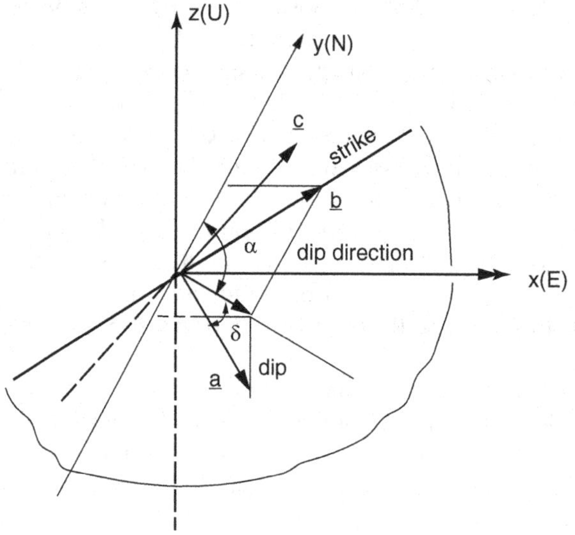

Figure 2.10 Fault plane coordinates (*abc*) from compass coordinates (*xyz*).

Table 2.1 Direction cosines between compass and joint plane coordinates

Compass axis Joint plane axis	x (east)	y (north)	z (up)
a (dip)	$\cos(\delta)\sin(\alpha)$	$\cos(\delta)\sin(\alpha)$	$-\sin(\delta)$
b (strike)	$-\cos(\alpha)$	$\sin(\alpha)$	0
c (normal)	$\sin(\delta)\sin(\alpha)$	$\sin(\delta)\cos(\alpha)$	$\cos(\delta)$

a = dip direction (+clockwise from north 0° to 360°).
o = down (+down 0°–90°).

Application of the vector dot and cross products shows that the new base vectors *a*, *b*, and *c* are indeed mutually orthogonal vectors of unit length. The direction cosines n_x, n_y, and n_z can now be computed from the third row of Table 2.1 with $a = 120°$ and $o = 60°$. Thus,

$$n_x = \sin(\delta)\sin(\alpha) = \left(\frac{\sqrt{3}}{2}\right)\left(\frac{\sqrt{3}}{2}\right) = \frac{3}{4}$$

$$n_y = \sin(\delta)\cos(\alpha) = \left(\frac{\sqrt{3}}{2}\right)\left(\frac{1}{2}\right) = \frac{\sqrt{3}}{4}$$

$$n_z = \cos(\delta) = \left(\frac{1}{2}\right)$$

According to (23), the tractions on the fault plane are

$$T_x = (-300)(3/4) + 0 + 0 = -225$$
$$T_y = 0 + (-300)(-\sqrt{3}/40 + 0) = +75\sqrt{3}$$
$$T_z = 0 + 0 + (-1500)(1/2) = -750$$

and according to (27), the normal stress is

$$\sigma = (-2225)(3.4) + (75\sqrt{3})(-\sqrt{3}/4) + (-750)(1/2) = T\cos(\theta) = T \cdot n = N$$

Thus, $\sigma = -600$ psi. Fault shear strength $\tau_f = (0.7)(600) = 420$ psi. According to (25) and (28), the fault shear stress is

$$\tau = \sqrt{T^2 - N^2} = \{[(-225)^2 + (75\sqrt{3})^2 + (-750)^2] - (-600)^2\}^{1/2}.$$

Thus, the magnitude of the fault plane shear stress is 520 psi. Because the shear stress exceeds the shear strength, frictional slip on the fault impends.

2.4   Equality of shear stresses

A careful review of the previous development reveals that only force equilibrium has been required. A natural question to ask then concerns the implications of moment equilibrium. Consider moment equilibrium of the small cube in Figure 2.8a by taking moments about the x-axis. Thus,

$$\sum M_x = 0 = (\tau_{yz})(\Delta x \Delta z)(\Delta y/2)(2) - (\tau_{yz})(\Delta x \Delta y)(\Delta z/2)(2) \tag{29}$$

where each term has the form (stress) (area) (moment arm) and the (2) accounts for shear on opposite faces. Similar equations can be written for the y- and z-directions. The result is that moment equilibrium requires equality of shear stresses in the sense that

$$\tau_{xy} = \tau_{yx}, \ \tau_{yz} = \tau_{zy}, \ \tau_{zx} = \tau_{xz} \tag{30}$$

This is not the most general case imaginable, but except in rather unusual circumstances (e.g., magnetic material in an intense magnetic field), the equality of shear stresses in (30) holds. As a consequence, only six rather than nine stresses are needed to specify a state of stress.

2.5   Rotation of reference axes

Rotation of the reference for stress (and strain) is an important aid to problem solving. A set of transformation formulas is needed for this purpose. Transformation of the components of a vector under a rotation of the reference axis amounts to simply projecting the components of the given vector onto the rotated axes. Magnitude or length of the vector is a scalar and remains unchanged. Transformation of stress (and strain) is more complicated. Stress is not a vector; not only do the components of stress change, but so does area under a rotation of reference axis.

A two-dimensional development shows the essential features of the transformation that is based on rotation of the reference axes and the definition of stress.

Consider a rotation of Cartesian reference axes from (x, y, z) to (n, s, t) about the z-axis through a positive counterclockwise angle a as shown in Figure 2.11. The rotation is

$$
\begin{aligned}
n &= x\cos(n,x) + y\cos(n,y) \\
s &= -x\cos(n,x) + y\cos(n,y) \\
t &= z
\end{aligned}
\tag{31}
$$

which can also be written as

$$
\begin{aligned}
n &= x\cos(\alpha) + y\sin(\alpha) \\
s &= -x\sin(\alpha) + y\cos(\alpha) \\
t &= z
\end{aligned}
\tag{32}
$$

By definition, stresses are the components of the stress vectors acting on areas parallel to the coordinate planes. These components can be calculated in two steps. First, the stress vector components relative to the old system are obtained from (23).

These components are then resolved onto the new coordinate directions. Both steps are shown in Figure 2.12a and b. The second step is

$$
\begin{aligned}
T_n^n &= T_x^n \cos(\alpha) + T_y^n \sin(\alpha) \\
T_s^n &= -T_x^n \sin(\alpha) + T_y^n \cos(\alpha)
\end{aligned}
$$

By definition, $\sigma_{nn} = T_n^n$ and $\tau_{ns} = T_s^n$. The process is now repeated for the s-axis and the coordinate plane perpendicular to it. Thus,

$$
\begin{aligned}
T_s^s &= T_x^n \cos(\alpha + \pi/2) + T_y^n \sin(\alpha + \pi/2) \\
T_n^s &= -T_x^n \sin(\alpha + \pi/2) + T_y^n \cos(\alpha + \pi/2)
\end{aligned}
$$

Again, by definition $\sigma_{ss} = T_s^s$ and $\tau_{sn} = T_n^s$. After substituting (22) into these results, one obtains the two-dimensional transformation formulas for stress under a rotation of the references axes. Thus,

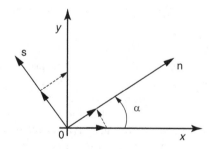

Figure 2.11 Rotation about the z-axis.

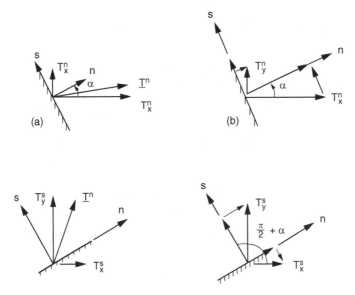

Figure 2.12 Rotation of reference axes about the z-axis.

$$\sigma_{nn} = \sigma_{xx}\cos^2(\alpha) + 2\tau_{xy}\sin(\alpha)\cos(\alpha) + \sigma_{yy}\sin^2(\alpha)$$
$$\tau_{ns} = -\sigma_{xx}\cos(\alpha)\sin(\alpha) + \tau_{xy}[\cos^2(\alpha) - \sin^2(\alpha)] + \sigma_{yy}\sin(\alpha)\cos(\alpha)$$
$$\tau_{sn} = -\sigma_{xx}\sin(\alpha)\cos(\alpha) + \tau_{xy}[\cos^2(\alpha) - \sin^2(\alpha)] + \sigma_{yy}\cos(\alpha)\sin(\alpha) \tag{33}$$
$$\sigma_{nn} = \sigma_{xx}\sin^2(\alpha) - 2\tau_{xy}\sin(\alpha)\cos(\alpha) + \sigma_{yy}\cos^2(\alpha)$$

which shows shear stresses τ_{ns} and τ_{sn} are equal as might be expected. Use of the identities $\sin^2(\alpha) + \cos^2(\alpha) = 1$ and $\cos^2(\alpha) - \sin^2(\alpha) = \cos(2\alpha)$ in (33) allows for some simplifications. Thus,

$$\sigma_{nn} = (1/2)(\sigma_{xx} + \sigma_{yy}) + (1/2)(\sigma_{xx} - \sigma_{yy})\cos(2\alpha) + \tau_{xy}\sin(2\alpha)$$
$$\sigma_{ss} = (1/2)(\sigma_{xx} + \sigma_{yy}) - (1/2)(\sigma_{xx} - \sigma_{yy})\cos(2\alpha) - \tau_{xy}\sin(2\alpha) \tag{34a}$$
$$\tau_{ns} = \tau_{sn} = -(1/2)(\sigma_{xx} - \sigma_{yy})\sin(2\alpha) + \tau_{xy}\cos(2\alpha)$$

If stresses σ_{zz}, τ_{zx}, and τ_{zy} act on the ends of the triangular prism in Figure 2.12, then because the triangular end area does not change, one has

$$\sigma_{tt} = \sigma_{zz}$$
$$\tau_{tn} = \tau_{zx}\cos(\alpha) + \tau_{zy}\sin(\alpha) \tag{34b}$$
$$\tau_{ts} = -\tau_{zx}\sin(\alpha) + \tau_{zy}\cos(\alpha)$$

Equations (34a and b) are the equations of transformation that relate the stresses given in the x, y, z system to the n, s, t system rotated about the z-axis.

A general rotation of axes in three dimensions may be derived using a similar, but lengthier procedure. The result in matrix form requires a 3×3 matrix of direction cosines $[R]$ that may be obtained from Table 2.1 with a change in notation such that

ax, ay, az are the first row entries in the table; b_x, b_y, b_z and c_x, c_y, c_z are the second and third row entries, respectively. Previously, the third row entries were indicated as n_x, n_y, n_z (normal direction). The rotation is then

$$
\begin{bmatrix}
\sigma_{aa} & \tau_{ba} & \tau_{ca} \\
\tau_{ab} & \sigma_{bb} & \tau_{cb} \\
\tau_{ac} & \tau_{bc} & \sigma_{cc}
\end{bmatrix}
= [R]
\begin{bmatrix}
\sigma_{xx} & \tau_{yx} & \tau_{zx} \\
\tau_{xy} & \sigma_{yy} & \tau_{zy} \\
\tau_{xz} & \tau_{yz} & \sigma_{zz}
\end{bmatrix}
[R]^T
\tag{34c}
$$

where the superscript T means transpose. The previous rotation about the z-axis has the matrix form

$$
\begin{bmatrix}
\sigma_{nn} & \tau_{sn} & \tau_{tn} \\
\tau_{ns} & \sigma_{ss} & \tau_{ts} \\
\tau_{nt} & \tau_{st} & \sigma_{tt}
\end{bmatrix}
=
\begin{bmatrix}
\cos(\alpha) & \sin(\alpha) & 0 \\
-\sin(\alpha) & \cos(\alpha) & 0 \\
0 & 0 & 1
\end{bmatrix}
[\sigma(xyz)][R]^T
$$

Example 2.2

Suppose the stress measurements with respect to x, y (x is east, and y is north) show that $\sigma_{xx} = -1,500$, $\tau_{xy} = 450$, $\sigma_{yy} = -300$, where tension is positive and the units are N/m^2. These data were obtained near a vertical joint plane that strikes N30W as shown in Figure 2.13. Let $\sigma_{nn} = \sigma$ be a normal stress and $\tau_{ns} = \tau$ be the shear stress on a joint plane relative to an n, s set of axes shown in Figure 2.13. According to the equations of transformation (34),

$$
\sigma_{nn} = (1/2)(-1,500 + -300) + (1/2)(-1,500 - -300)\cos(2 \cdot 30°) + 450\sin(2 \cdot 30°)
$$
$$
\tau_{ns} = (1/2)(-1,500 - -300)\sin(2 \cdot 30°) + 450\cos(2 \cdot 30°)
$$

Hence, $\sigma = -810$ N/m^2 (compressive) and $\tau = 745$ N/m^2 are the normal and shear stresses acting on the joint. The normal stress σ_{ss} acting parallel to the joint (but on a surface at right angles to the joint) is -990 N/m^2, which is also obtained from (34).

2.6   Principal stresses

Inspection of the two-dimensional equations of transformation shows that the normal and shear stresses depend on the angle of rotation α at any given point. A natural question to ask is whether there are directions in which the normal stress has a maximum or minimum value and whether the same is true of the shear stress. Practical motivation for such an inquiry arises in association with failure of materials. For example, if the failure criterion in tension is one that states failure occurs when the tension reaches a critical value, the material strength, then it is important to know what the maximum normal stress is in comparison with strength.

   The behavior of a material, although considered in reference to a coordinate system, is intrinsic to the material and must therefore be independent of the coordinate system. Failure should not depend on the orientation of the coordinate system that is

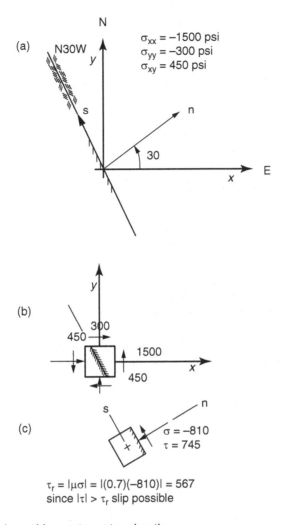

N

(a) N30W

$\sigma_{xx} = -1500$ psi
$\sigma_{yy} = -300$ psi
$\sigma_{xy} = 450$ psi

y

s

n

30

E

x

(b)

y

300
450

1500

x

450

(c)

s

n

$\sigma = -810$
$\tau = 745$

$\tau_r = |\mu\sigma| = |(0.7)(-810)| = 567$
since $|\tau| > \tau_r$ slip possible

Figure 2.13 Example problem orientation details.

arbitrarily established for problem analysis. Thus, in seeking directions that the normal and shear stresses assume extreme values, one is not only exploring the mathematical aspects of stress, but also acquiring information important to design analysis.

Because the point of interest is fixed for this analysis, the stresses vary only with direction. According to (34), the normal stress acting on a surface with a normal inclined at an angle a to the x-axis is

$$\sigma = (1/2)(\sigma_{xx} + \sigma_{yy}) + (1/2)(\sigma_{xx} - \sigma_{yy})\cos(2\alpha) + \tau_{xy}\sin(2\alpha) \qquad (35)$$

where the change in notation $\sigma = \sigma_{nn}$ has been made. A necessary condition

$$d\sigma/d\alpha = 0 = -2(1/2)(\sigma_{xx} - \sigma_{yy})\sin(2\alpha) + 2\tau_{xy}\cos(2\alpha)$$

that σ has a stationary value at the considered point is

$$\tan(2\alpha^*) = \frac{\tau_{xy}}{(1/2)(\sigma_{xx} - \sigma_{yy})} \tag{36}$$

At values of $\alpha = \alpha^*$, the normal stress is stationary. From (36),

$$\sin(2\alpha^*) = \frac{\tau_{xy}}{\pm\left[\left(\frac{\sigma_{xx} - \sigma_{yy}}{2}\right)^2 + (\tau_{xy})^2\right]^{1/2}}$$

$$\cos(2\alpha^*) = \frac{\left(\frac{\sigma_{xx} - \sigma_{yy}}{2}\right)}{\pm\left[\left(\frac{\sigma_{xx} - \sigma_{yy}}{2}\right)^2 + (\tau_{xy})^2\right]^{1/2}}$$

The second derivative

$$d^2\sigma/d\alpha^2 = -4(1/2)(\sigma_{xx} - \sigma_{yy})\sin(2\alpha) - 4\tau_{xy}\cos(2\alpha)$$

is negative when the plus sign is used for sine and cosine and positive when the minus sign is used. The first corresponds to a maximum, and the second to a minimum value of the normal stress. If σ_1 and σ_3 denote the *algebraic* maximum and minimum values of σ, then

$$\begin{Bmatrix} \sigma_1 \\ \sigma_3 \end{Bmatrix} = \left(\frac{\sigma_{xx} + \sigma_{yy}}{2}\right) \pm \left[\left(\frac{\sigma_{xx} - \sigma_{yy}}{2}\right)^2 + (\tau_{xy})^2\right]^{1/2} \tag{37}$$

which is obtained by substitution of cosine and sine values into (35). The stresses σ_1 and σ_3 are *principal stresses*: σ_1 is the major principal stress, and σ_3 is the minor principal stress. Equation (36) shows that the major and minor principal stresses act on surfaces at right angles to each other. If α_1 and α_3 are the angles from the x-axis to the normals to the surfaces acted on by σ_1 and σ_3, then $2\alpha_3 = \pi + 2\alpha_1$ and therefore $\alpha_3 = \pi/2 + \alpha_1$. The proper quadrant for σ_1 can be determined by the combination of algebraic signs of $\sin(2\alpha_1)$ and $\cos(2\alpha_1)$. The direction of action of σ_1 is in all cases nearest to the algebraically larger of σ_{xx} or σ_{yy}, as might be expected on physical grounds.

The planes on which the principal stresses σ_1 and σ_3 act are *principal planes*. Principal planes have the interesting property of being free of shear stress. To see that this is the case, one needs only to calculate the shear stresses τ_{ns} and τ_{sn} (for both values of α). The converse is also true; a surface free of shear stress must be a principal plane.

From point to point, the direction of principal stress and, therefore, the orientation of the principal planes generally change. In large, principal surfaces are generally curved. Nevertheless, if a surface is free of shear stress, it is a principal surface and one of the principal stresses acts perpendicular to it. The other principal stress acts parallel to this surface.

Principal surfaces are quite common. In fact, all unsupported excavation surfaces must be principal surfaces because they are free of shear stress (and normal stress as well). This observation gives important information about unsupported excavations without the need to do a complicated analysis of the stress field in the adjacent rock mass. Because the normal stress acting perpendicular to the excavation surface is zero and the excavation surface is a principal surface, the stress of importance to the stability of the opening must be a principal stress acting in a direction parallel to the excavation (on interior surfaces perpendicular to the surface of the excavation) as illustrated in Figure 2.14.

Example 2.3

Suppose the stresses at a point of interest are $\sigma_{xx} = 300$, $\tau_{xy} = -1,200$ and $\sigma_{yy} = -700$ psi and compression is positive as shown in Figure 2.15a. The problem is to determine the principal stresses and directions. According to (37), the principal stresses are

$$\left\{ \begin{matrix} \sigma_1 \\ \sigma_3 \end{matrix} \right\} = \left(\frac{300 + -700}{2} \right) \pm \left[\left(\frac{300 - -700}{2} \right)^2 + (-1,200)^2 \right]^{1/2} = \left\{ \begin{matrix} 1,100 \\ -1,200 \end{matrix} \right\} \text{ psi}$$

The direction of σ_1 relative to the x-axis is obtained from

$$\tan(2\alpha^*) = \frac{-1,200}{(1/2)\,(300 - -700)} = -2.4$$

There are two values of $2\alpha^*$: $2\alpha^* = -67.4°$ and $2\alpha^* = +112.6°$. However, the sine is negative and the cosine is positive, so the proper choice is $2\alpha^* = -67.4°$; hence, $\alpha^* = \alpha_1 = -33.7°$, as shown in Figure 2.15b.

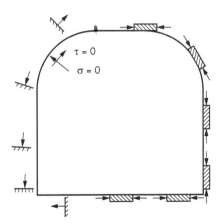

Figure 2.14 Unsupported excavation surfaces are traction-free and thus are principal surfaces.

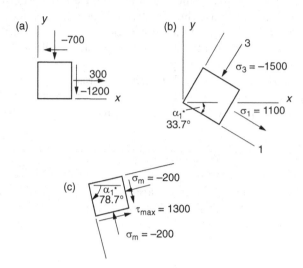

Figure 2.15 Geometry and numerical results for Example 2.3.

Vanishing of shear stress on the principal planes can be verified by substituting $\alpha_1 = -33.7°$ and $\alpha_1 = -33.7° + 90° = 56.3°$ into the last expression of (34). Thus to within round off accuracy

$$\tau = -\left(\frac{300+700}{2}\right)\sin\left(\frac{-67.4°}{112.6°}\right) + (91,200)\cos\left(\frac{-67.4°}{112.6°}\right)$$
$$\tau = \pm462 \mp 462 = 0$$

2.7 Maximum and minimum shear stresses

The maximum shearing stress is found by the same procedure used to find the maximum normal stress (major principal stress). Differentiating the last expression of (34), one obtains

$$d\tau/d\alpha = -(\sigma_{xx} - \sigma_{yy})\cos(2\alpha) - 2\tau_{xy}\sin(2\alpha)$$

The critical values of a are then given by the solutions to

$$\tan(2\alpha **) = -(1/2)(\sigma_{xx} - \sigma_{yy})/\tau_{xy} \tag{38}$$

Comparison of (38) with (36) shows that the product of the slopes $\tan(2\alpha*)\tan(2\alpha **)$ equals a minus one. The directions specified by $2\alpha*$ and $2\alpha **$ are thus at right angles, so that the directions of $\alpha*$ and $\alpha **$ are at 45° to one another. There are two values of $\alpha*$, of course, and two values of $\alpha**$ corresponding to algebraic maximum and minimum shear stresses. These are $\alpha_1 ** = \alpha_1 * -\pi/4$ and $\alpha_3 ** = \alpha_1 * +\pi/4$. The directions of maximum shear and minimum shear stress and the directions of the principal stresses are shown in Figure 2.16.

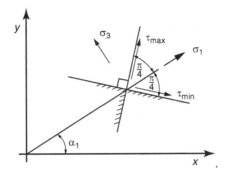

Figure 2.16 Orientation of maximum and minimum shear stresses with respect to principal axes.

The values of the maximum and minimum shear stress, τ_{max} and τ_{min} and the fact that they are indeed maximum and minimum, can be determined by back-substitution of the critical angles a^{**} into the last expression of (34). The result is

$$\begin{Bmatrix} \tau_{max} \\ \tau_{min} \end{Bmatrix} = \pm \left[\left(\frac{\sigma_{xx} - \sigma_{yy}}{2} \right)^2 + (\tau_{xy})^2 \right]^{1/2} \tag{39}$$

The only difference between the maximum and minimum shear stresses at a considered point is one of algebraic sign and thus of direction; their magnitudes are the same. One may note that the maximum shear stress is just one half the difference between the principal stresses, that is,

$$\tau_{max} = (1/2)\,(\sigma_1 - \sigma_3)$$

as can be readily seen from (37) and (39).

Normal stress does not vanish on the planes of maximum and minimum shear stress, but rather is given by the two-dimensional mean normal stress σ_m that can be calculated using the equations of transformation and the values of α^{**}. The result is
$\sigma_m = (1/2)\,(\sigma_{xx} + \sigma_{yy})$.

Example 2.4

The value of τ_{max} for the stress state given in Example 1.4 and according to (40) is

$$\tau_{max} = (0.5)\,(1{,}100 - -1{,}500) = 1{,}300 \text{ psi}$$

Alternatively,

$$\tau_{max} = \left[(1/209{,}300 - -700)^2 + (-1{,}200)^2 \right]^{1/2} = 1{,}300 \text{ psi}$$

from (39). The direction of the normal to the plane acted on by τ_{max} is given by (38); thus,

$$\tan(2\alpha **) = (-0.5)(300 - -700)/(-1,200) = 0.4167$$

The two values of $2\alpha **$ given by (38) are 22.6° and 202.6°. According to the formulas

$$\cos(2\alpha **) = \frac{\tau_{xy}}{\pm\left[\left(\dfrac{\sigma_{xx} - \sigma_{yy}}{2}\right)^2 + (\tau_{xy})^2\right]^{1/2}}$$

$$\sin(2\alpha **) = \frac{\left(\dfrac{\sigma_{xx} - \sigma_{yy}}{2}\right)}{\pm\left[\left(\dfrac{\sigma_{xx} - \sigma_{yy}}{2}\right)^2 + (\tau_{xy})^2\right]^{1/2}} \tag{41}$$

$2\alpha_1 **$ must be in the third quadrant because both the cosine and sine are negative. Hence, the value of $2\alpha_1 **$ is 101.3° equivalently, −78.7° as shown in Figure 2.15c. As a check, one notes that $\alpha_1 ** = \alpha_1 * - \pi/4$ (−78.7 = −33.7−45). The normal stress acting on the planes of maximum and minimum shear stress is $\sigma_m = (0.5)(300 + -700) = -200$ psi.

2.8   Stress invariants, hydrostatic stress, and deviatoric stress

Invariants are quantities that remain unchanged under a rotation of the reference axes. Such quantities are important to the description of material strength that should not depend on the choice of reference axes. Values of the two-dimensional mean normal stress σ_m and the maximum shear stress $\tau_{\max}$ are important invariants.

Addition of (37) and separately the first two of (34) shows that stresses $(\sigma_1 + \sigma_3) = (\sigma_{xx} + \sigma_{yy}) = (\sigma_{nn} + \sigma_{ss})$, and since orientation of the x- and y-axes is arbitrary, the mean normal stress $\sigma_m = (\sigma_{xx} + \sigma_{yy})/2$ is indeed unchanged when the state of stress is referred to a rotated set of axes. In fact, if the reference axes are chosen to coincide with the principal directions, then

$$\begin{aligned}
\sigma_{xx} &= (1/2)(\sigma_1 + \sigma_3) + (1/2)(\sigma_1 - \sigma_3)\cos(2\alpha_1) \\
\sigma_{yy} &= (1/2)(\sigma_1 + \sigma_3) - (1/2)(\sigma_1 - \sigma_3)\cos(2\alpha_1) \\
\tau_{xy} &= \tau_{yx} = -(1/2)(\sigma_1 - \sigma_3)\sin(2\alpha_1)
\end{aligned} \tag{42}$$

where a is the angle between the 1-axis and the x-axis and, as always, counterclockwise angles are positive, as shown in Figure 2.17. The absence of shear stresses in (42) reflects the fact that principal planes are shear-free. A second stress invariant is $\sigma_{xx}\sigma_{yy} - \tau_{xy}\tau_{yx}$ the product of the principal stresses $\sigma_1\sigma_3$; there are no other basic invariants of stress in two dimensions.

Deviatoric normal stresses are by definition the departure of the normal stresses from the mean normal stress that is also referred to as the hydrostatic stress. Deviatoric shear stresses are just the shear stresses proper. In two dimensions, the deviatoric stresses relative to x, y are

$$\begin{aligned}
s_{xx} &= \sigma_{xx} - (1/2)(\sigma_{xx} + \sigma_{yy}) \\
s_{yy} &= \sigma_{yy} - (1/2)(\sigma_{xx} + \sigma_{yy}) \\
s_{xy} &= \tau_{xy}
\end{aligned} \tag{43}$$

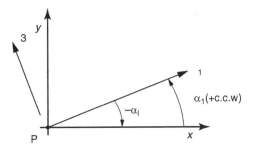

Figure 2.17 Rotation of reference axes between principal axes and Cartesian axes.

In three dimensions, the mean normal stress is the average of all three normal stresses. Thus, $\sigma_m = (1/3)(\sigma_{xx} + \sigma_{yy} + \sigma_{zz})$, which is also known as the (three-dimensional) hydrostatic stress. The deviatoric stresses in three dimensions are

$$
\begin{aligned}
s_{xx} &= \sigma_{xx} - (1/3)(\sigma_{xx} + \sigma_{yy} + \sigma_{zz}), \quad s_{xy} = \tau_{xy} \\
s_{yy} &= \sigma_{yy} - (1/3)(\sigma_{xx} + \sigma_{yy} + \sigma_{zz}), \quad s_{yz} = \tau_{yz} \\
s_{zz} &= \sigma_{zz} - (1/3)(\sigma_{xx} + \sigma_{yy} + \sigma_{zz}), \quad s_{zx} = \tau_{zx}
\end{aligned}
\tag{44}
$$

Inspection of (43) and (44) shows that the sum of the deviatoric stresses (mean deviatoric stress) is zero. A more interesting quantity in two dimensions is the second deviatoric invariant. This quantity can also be written as $-\left[(1/4)(\sigma_{xx} - \sigma_{yy})^2 + (\tau_{xy})^2\right]$ or as $-\left[(1/4)(\sigma_{nn} - \sigma_{ss})^2 + (\tau_{ns})^2\right]$ or even as $-\left[(1/4)(\sigma_1 - \sigma_1)^2\right]$. This last expression shows that the maximum (and minimum) shear stresses are also invariants. In three dimensions, the second invariant of deviatoric stress is the more complicated expression $-\left[(1/4)(\sigma_{xx} - \sigma_{yy})^2 + (\tau_{xy})^2 + (1/4)(\sigma_{yy} - \sigma_{zz})^2 + (\tau_{yz})^2 + (1/4)(\sigma_{zz} - \sigma_{xx})^2 + (\tau_{zx})^2\right]$ and is quite useful in formulating the failure criteria. This expression has the alternative $-\left[(1/4)(\sigma_1 - \sigma_2)^2 + (1/4)(\sigma_2 - \sigma_3)^2 + (1/4)(\sigma_3 - \sigma_1)^2\right]$, where $\sigma_1, \sigma_2, \sigma_3$ are the algebraic major, *intermediate*, and minor principal stresses. Principal stresses act in mutually orthogonal directions. The principal stress differences divided by 2 are relative maxima and minima shear stresses $\pm(1/2)(\sigma_1 - \sigma_2), \pm(1/2)(\sigma_2 - \sigma_3), \pm+(1/2)(\sigma_3 - \sigma_1)$. Derivations of the three-dimensional relations are rather lengthy and so are simply stated here for future reference. Deviatoric stresses also have principal values and directions as well as invariants, and they transform under rotation of the reference axes in the same way that the stresses transform.

2.9   Mohr's circle in two dimensions

A graphical representation of the two-dimensional equations of stress transformation provides a convenient method for representing the state of stress at a point and for a quick check on calculations. The technique is known as Mohr's circle. The equation of a circle of radius R centered on the x-axis at $(a, 0)$ is

$$
(x - a)^2 + y^2 = R^2
\tag{45}
$$

as illustrated in Figure 2.18a. Points (x, y) that satisfy (45) lie on the perimeter of the circle. A similar equation can be derived in terms of normal and shear stresses. Let $\sigma = \sigma_m$ and $\tau = \tau_m$. With this change, one has from the first and last expressions of (34a)

$$\sigma - (1/2)(\sigma_{xx} + \sigma_{yy}) = +(1/2)(\sigma_{xx} - \sigma_{yy})\cos(2\alpha) + \tau_{xy}\sin(2\alpha)$$
$$\tau = -(1/2)(\sigma_{xx} - \sigma_{yy})\sin(2\alpha) + \tau_{xy}\cos(2\alpha)$$

After squaring and adding, one obtains

$$\left[\sigma - (1/2)(\sigma_{xx} + \sigma_{yy})\right]^2 + \tau^2 = \left[(1/2)(\sigma_{xx} - \sigma_{yy})\right]^2 + (\tau_{xy})^2 \tag{46a}$$

$$[\sigma - \sigma_m]^2 + \tau^2 = (\tau_m)^2 \tag{46b}$$

where the definitions of mean normal stress and maximum shear stress σ_m and τ_m are used in (46a) to obtain (47b). Both describe a circle in the σ, τ (normal stress, shear stress) plane centered at σ_m with radius τ_m as shown in Figure 2.18b. To find

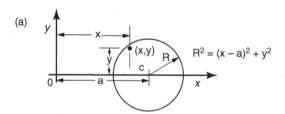

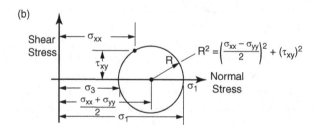

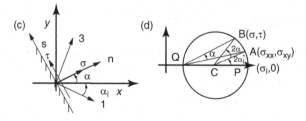

Figure 2.18 Geometry of Mohr's circle.

the stresses σ, τ acting on a plane with a normal inclined at an angle α to the x-axis, one can use the equations of transformation, the first and last expressions of (34), or Mohr's circle. The latter can be confusing because of the sign convention and angular relationships. Figure 2.18c and d shows the relationship between the physical plane and the stress plane containing Mohr's circle. In Figure 2.18d, the given information consists of stresses $\sigma_{xx}, \sigma_{yy}, \tau_{xy}$. From these data, one can compute the corresponding center and radius of Mohr's circle. The angle from the x-axis to the direction of σ_1 is α_1 and is determined graphically by measuring the angle AQP in Figure 2.18d. The normal and shear stresses σ, τ acting on some other plane with a normal inclined at an angle α to the x-axis are determined by laying out angle AQB equal to α. The abscissa of the point B formed by intersection of a line drawn from Q at an angle α to AQ is σ; the ordinate is τ. In three dimensions, the state of stress can be represented by three Mohr's circles. The two additional circles are contained within the largest Mohr circle that has a radius $R = (1/2)(\sigma_1 - \sigma_3)$.

2.10   Stress equations of motion and equilibrium

Investigation of the force side of the balance law for linear momentum led to the concept of stress and the use of stress to characterize the internal mechanical reaction of a body to externally applied loads. Investigation of the momentum side of the balance remains. This investigation is quite direct and utilizes the previously developed concept of stress to calculate the momentum of a typical mass element within a body. Summation of the contributions of all mass elements in the body completes the calculation. The latter step is simply integration. However, the first step requires a detailed examination of a typical mass element dm within the body of interest. The immediate outcome is a stress system of equations of motion or equilibrium, in case of vanishing accelerations. An interesting subsidiary result is a relationship between the resultant of external forces and the stresses in the interior of the body. All the essential features of the investigation are contained in a two-dimensional analysis.

To begin, subdivide the body of interest into a number of elements dm that occupy volume dV and isolate a typical element from the body for a detailed examination, as shown in Figure 2.19.

The effect of the material removed on the material remaining, the element dm, is represented by distributed surface and body forces. The first act on the faces of the considered element and are tractions; the components are by definition the stresses because the faces of the element are parallel to the coordinate planes. In general, the stresses vary from point to point. If the stress at x, y is σ_{xx}, then the stress at $(x+dx, y)$ is $\sigma_{xx}(x+dx, y) = \sigma_{xx}(x, y) + \Delta\sigma_{xx}$, where $\Delta\sigma_{xx}$ is the change between the left- and right-hand sides of the element and is given by $\Delta\sigma_{xx} = (\partial\sigma_{xx}/\partial x)(dx)$.

Strictly speaking, some allowance should be made for the variation over the height of the element sides. However, as the dimensions of the element become small such effects vanish. For the remaining stresses, one obtains

$$\sigma_{yy}(x, y+dy) = \sigma_{yy}(x, y) + \Delta\sigma_{yy}, \quad \Delta\sigma_{yy} = (\partial\sigma_{yy}/\partial y)dy$$
$$\tau_{xy}(x+dx, y) = \tau_{xy}(x, y) + \Delta\tau_{xy}, \quad \Delta\tau_{yy} = (\partial\tau_{xy}/\partial x)dx$$
$$\tau_{yx}(x, y+dy) = \tau_{yx}(x, y) + \Delta\tau_{yx}, \quad \Delta\tau_{yx} = (\partial\tau_{yx}/\partial y)dy$$

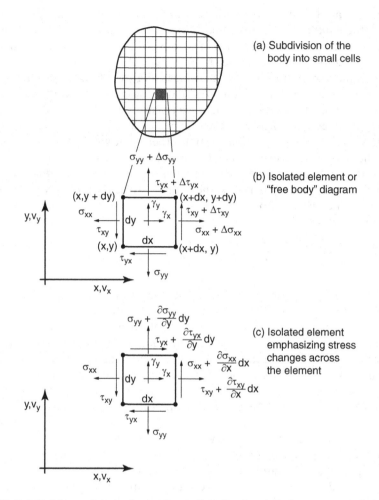

(a) Subdivision of the
body into small cells

(b) Isolated element or
"free body" diagram

(c) Isolated element
emphasizing stress
changes across
the element

Figure 2.19 Subdivision of a body into small cells for balancing internal forces and accelerations.

The x- and y-components of the time rate of change of linear momentum are

$$\dot{v}_x dm = a_x dm, \ \dot{v}_y dm = a_y dm$$

where v is velocity and a is acceleration. Summing forces in the x-direction, one obtains

$$(\sigma_{xx} + \Delta\sigma_{xx})dy - (\sigma_{xx})dy + (\tau_{yx} + \Delta\tau_{yx})dx - (\tau_{yx})dx + \gamma_x dydx = a_x dm$$

where a unit thickness into the plane of the page is implied. Substituting $dV = dxdy$ and $\rho dV = dm$,

$$(\partial\sigma_{xx}/\partial x)dxdy + (\partial\tau_{yx}/\partial y)dxdy + \gamma_x dxdy = \rho a_x dxdy$$

Integrating over the volume of the body results in

$$\int_V \left(\frac{\partial \sigma_{xx}}{\partial x} + \frac{\partial \tau_{yx}}{\partial y} + \gamma_x \right) dV = \int_V \rho a_x \, dV \tag{47a}$$

A similar process for the y-direction leads to

$$\int_V \left(\frac{\partial \tau_{xy}}{\partial x} + \frac{\partial \sigma_{yy}}{\partial y} + \gamma_y \right) dV = \int_V \rho a_y \, dV \tag{47b}$$

Equations (47) are the two-dimensional stress equations of motion in integral form. If one brings the right-hand sides under the integrals on the left and notes that the result must hold for arbitrary volumes, one obtains these equations in differential form. Thus,

$$\begin{aligned}
\frac{\partial \sigma_{xx}}{\partial x} + \frac{\partial \tau_{yx}}{\partial y} + \gamma_x &= \rho a_x \\
\frac{\partial \tau_{xy}}{\partial x} + \frac{\partial \sigma_{yy}}{\partial y} + \gamma_y &= \rho a_y
\end{aligned} \tag{48}$$

If the accelerations are negligible, then one has the stress equations of equilibrium. In three dimensions, the stress equations of motion are

$$\begin{aligned}
\frac{\partial \sigma_{xx}}{\partial x} + \frac{\partial \tau_{yx}}{\partial y} + \frac{\partial \tau_{zx}}{\partial z} + \gamma_x &= \rho a_x \\
\frac{\partial \tau_{xy}}{\partial x} + \frac{\partial \sigma_{yy}}{\partial y} + \frac{\partial \tau_{xz}}{\partial z} + \gamma_y &= \rho a_y \\
\frac{\partial \tau_{xz}}{\partial x} + \frac{\partial \tau_{yz}}{\partial y} + \frac{\partial \sigma_{zz}}{\partial z} + \gamma_z &= \rho a_z
\end{aligned} \tag{49}$$

Equations (48) constitute a system of two equations involving six unknowns $\sigma_{xx}, \sigma_{yy}, \tau_{xy}, \rho, a_x, a_y$. The conservation of mass adds a third equation to the system, but also introduces velocities, v_x, v_y, but since the accelerations are time derivatives of velocity, in effect, no new unknowns are introduced. Thus, in two dimensions there is a deficit of equations; three additional independent equations are needed to complete the system. In three dimensions, a count reveals ten unknowns (six stresses, density, and three accelerations), but there are only four equations. Thus, there is a deficit of six equations in three dimensions. Hence, physical laws (mass conservation, balance of linear momentum, and balance of angular momentum) are not sufficient to determine the motion of a deformable body. Application of the balance law for angular momentum (moment of momentum) does not produce additional information about the stress equations of motion, other than equality of shear stresses, and does not therefore restrict them in any way.

An interesting result of the preceding investigation arises from comparison of (47) with (9) that shows that the right-hand sides are identical. Thus, the left-hand sides must be equal. Hence, in two dimensions

$$\int_S T_x \, dS = \int_S (\sigma_{xx} n_x + \tau_{yx} n_y) \, dS = \int_V \left(\frac{\partial \sigma_{xx}}{\partial x} + \frac{\partial \tau_{yx}}{\partial y} \right) dV$$

$$\int_S T_y \, dS = \int_S (\tau_{xy} n_x + \sigma_{yy} n_y) \, dS = \int_V \left(\frac{\partial \tau_{xy}}{\partial x} + \frac{\partial \sigma_{yy}}{\partial y} \right) dV \tag{50}$$

Equation (50) and the three-dimensional version is nothing more than a statement of the well-known divergence theorem that transforms a surface integral into a volume integral.

2.11   Initial stress and stress change

The majority of problems in rock mechanics are associated with rock masses that are stressed prior to excavation. The reason is that rock *in situ* is load by gravity and possibly by forces of tectonic origin prior to any excavation. Excavation induces stress changes. The final state of stress is the sum of the initial stress and the stress change. The speed of a stress change leads to the concept of stress rate.

If σ^o represents the initial state of stress and is the stress change or increment, then the final state of stress σ is simply $\sigma = \sigma^o + \Delta\sigma$. In general,

$$\sigma_{xx} = \sigma_{xx}^o + \Delta\sigma_{xx}, \ \tau_{xy} = \tau_{xy}^o + \Delta\tau_{xy}$$
$$\sigma_{yy} = \sigma_{yy}^o + \Delta\sigma_{yy}, \ \tau_{yz} = \tau_{yz}^o + \Delta\tau_{yz} \tag{51}$$
$$\sigma_{zz} = \sigma_{zz}^o + \Delta\sigma_{zz}, \ \tau_{zx} = \tau_{zx}^o + \Delta\tau_{zx}$$

The increments are just the differences between the final and initial stresses:

$$\Delta\sigma_{xx} = \sigma_{xx} - \sigma_{xx}^o, \ \Delta\tau_{xy} = \tau_{xy} - \tau_{xy}^o$$
$$\Delta\sigma_{yy} = \sigma_{yy} - \sigma_{yy}^o, \ \Delta\tau_{yz} = \tau_{yz} - \tau_{yz}^o$$
$$\Delta\sigma_{zz} = \sigma_{zz} - \sigma_{zz}^o, \ \Delta\tau_{zx} = \tau_{zx} - \tau_{zx}^o$$

Stress increments also have principal values and directions where the increments acquire maximum and minimum values. The mathematical derivation is formally the same as that for stress.

$$\left\{ \begin{matrix} \Delta\sigma_1 \\ \Delta\sigma_3 \end{matrix} \right\} = \left(\frac{\Delta\sigma_{xx} + \Delta\sigma_{yy}}{2} \right) \pm \left[\left(\frac{\Delta\sigma_{xx} - \Delta\sigma_{yy}}{2} \right)^2 + (\Delta\tau_{xy})^2 \right]^{1/2} \tag{52}$$

No increments of shear stress occur on the principal planes of the stress increments. Directions of the principal stress increments β_1 and β_3 can be obtained from

$$\tan(2\beta) = \frac{\Delta\tau_{xy}}{(1/2)(\Delta\sigma_{xx} - \Delta\sigma_{yy})} \tag{53}$$

Generally, the principal directions in the final state will be different from the initial state; that is, generally, $\alpha_1 \neq \alpha_1^o \neq \beta_1$. Because of the change in directions, it makes no sense to write

$$\Delta\sigma_1 = \Delta\sigma_1 - \Delta\sigma_1^o \quad \text{or} \quad \Delta\sigma_3 = \Delta\sigma_3 - \Delta\sigma_3^o \tag{54}$$

Changes in stress must be computed relative to the same *fixed* basis unless the change in reference is taken into account. Figure 2.20 illustrates the different principal directions.

There is no limit to the size of the stress increments or changes in (51), nor on the time that elapses as the changes occur. However, if an element of material originally at some place P^o under the influence of initial stresses moves a significant distance to some other place P, then a decision must be made as to whether one is interested in the stresses and stress changes that occur in the material element ("particle") or at a specified point in space. Usually, it is the former that is of interest. Fortunately, the difference between particle and place is negligible in many rock engineering problems. In geological applications, this difference is less likely to be negligible.

A stress rate is essentially a change of stress per unit time. However, there are different ways of computing stress rates that depend on one's interest to a certain extent and on the mechanics of the situation. For example, one may wish to compute a stress rate at a given point in space or a stress rate experienced by a given material element. One may also wish to compensate for changes in stress that occur during rigid body motion and any deformation as well. The various rates that evolve from such specifications are rather complicated; only the first two cases are given here.

In the case of stress change per unit time at a given point, the rate is simply

$$\dot{\sigma} = \frac{\partial\sigma(x,y,z,t)}{\partial t}\bigg|_{x,y,z} \tag{55}$$

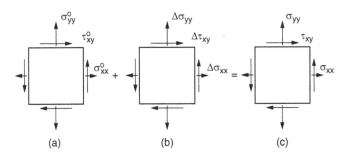

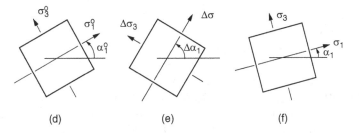

Figure 2.20 Principal directions of stress and stress change.

where σ stands for any stress component including shear stresses. A material element currently occupying the point (x, y, z) experiences the rate given by (55) that generally changes with time.

If instead of focusing attention on a point fixed in space, one focuses on a specified material element and wishes to record the time rate of stress change experienced by the element, one must first label the element. This task is easily done by specifying the coordinates of the element at the start, time t_o. At this time, $x = a$, $y = b$, $z = c$, and the stress acting on the element is $\sigma_e(a, b, c, t_o)$; at some later time t, the considered element is at x, y, z and the stress is $\sigma_e(x, y, z, t)$. The element experiences the rate formed by the limit

$$\lim_{t \to t_o} \left(\frac{\sigma_e(x, y, x, t) - \sigma_e(a, b, c, t_o)}{t - t_o} \right) = \frac{D\sigma_e}{Dt}$$

which is a stress rate "following the particle". This rate may be computed using the chain rule for differentiation after noting that the coordinates are functions of time. Thus,

$$\frac{D\sigma}{Dt} = \frac{\partial\sigma}{\partial t} + \frac{\partial\sigma}{\partial x}\frac{dx}{dt} + \frac{\partial\sigma}{\partial y}\frac{dy}{dt} + \frac{\partial\sigma}{\partial z}\frac{dz}{dt}$$

where the subscript e is no longer needed and again σ stands for any stress component of interest including shear stress. The ordinary derivatives are just velocities of the material point, so

$$\frac{D\sigma}{Dt} = \frac{\partial\sigma}{\partial t} + \frac{\partial\sigma}{\partial x}v_x + \frac{\partial\sigma}{\partial y}v_y + \frac{\partial\sigma}{\partial z}v_z \tag{56}$$

The two parts of (56) show a change at a fixed point ("local" part) and a contribution from movement, a "convective" part. If the velocities are zero, the element does not move to another point and the convective part vanishes. Quite often, the convective part is negligible compared with the local part. Indeed, the approximation $D\sigma/Dt = \partial\sigma/\partial t$ is widely used to describe dynamic wave phenomena.

3   ANALYSIS OF STRAIN

Application of external loads to a deformable body results in translation and rotation of the body as a whole and changes in size and shape of the body. Translation and rotation are shared by all points in the body as though the body were rigid. Size and shape changes involve deformation, that is, displacement of points relative to other points within the body. The concept of strain is intimately associated with the latter, that is, with relative displacement. There is no reason to suppose that under a given set of loads, all parts of a body experience identical changes in size and shape. In fact, even under controlled conditions in the laboratory, it is a difficult task to obtain a uniform state of deformation in a test specimen. Development of the concept of strain is therefore local rather than global in character and concerns deformation of a small generic element of material. Focus is on relative displacements that neighboring points experience in response to applied loads. Although strain is physically a very different concept from stress, the mathematical character of strain turns out to be very much the same.

3.1   Normal strain

The most common definition of normal strain is a change in length of a material element per unit of original length. Consider the small material element shown in Figure 3.1a. If the original length of the element measured along the x-axis is L_o and the present length is L, the normal strain ε, by definition, is

$$\varepsilon = \frac{L - L_o}{L_o} \tag{57}$$

as shown in Figure 3.1b. The same line of reasoning could, of course, be applied to the element of interest in the y- and z-directions, so there is immediately a need to distinguish between these three possible normal strains. The same notation used to distinguish normal stresses serves this purpose quite well. Thus,

ε_{xx} = strain associated with a surface element:

    (1) with an outward normal pointing in the x-direction,
    (2) acting in the x-direction

ε_{yy} = y-direction normal strain
ε_{zz} = z-direction normal strain

as illustrated in Figure 3.1c.

3.2   Shear strain

Shear strain refers to a change in the orientation of a line element in contrast to a change in length associated with normal strain. With reference to Figure 3.1d, a convenient way of describing the change in orientation would be to specify the angle γ. For small angles, $\gamma = \tan(\gamma)$, so that

$$\gamma = \frac{\Delta}{L_o} \tag{58}$$

Again, there are three possibilities for shear strain and thus a need to invoke a notation that distinguishes among them, and again, the same notation used for shear stresses is used for shear strains. Thus,

γ_{xy} = strain associated with a surface element:

    (1) with an outward normal pointing in the x-direction,
    (2) acting in the y-direction

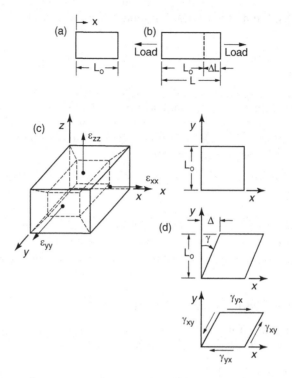

Figure 3.1 Geometry of primitive definitions of normal and shear strains.

γ_{yz} = z-direction normal strain
γ_{zx} = x-direction normal strain.

These shear strains are known as *engineering* shear strains. A different symbol γ is used for this distinction. Mathematically more convenient shear strains are just one half the engineering shear strains. The mathematical shear strains in double-subscript notation are

$$\varepsilon_{xy} = \gamma_{xy}/2, \quad \varepsilon_{yz} = \gamma_{yz}/2 \quad \varepsilon_{zx} = \gamma_{zx}/2 \tag{59}$$

When mathematical shear strains are used, it turns out $\varepsilon_{xx}, \varepsilon_{yy}, \varepsilon_{zz}, \varepsilon_{xy}, \varepsilon_{yz}, \varepsilon_{zx}$, formally similar to stress. What has been said about principal stress, maximum and minimum shear stresses, rotation of reference axes, and so forth also applies to strain in two and three dimensions.

3.3 Small strain–displacement relations

If one attempts to apply the definitions of normal and shear strains given by (57) and (58) directly, unforeseen complications arise that force a more precise treatment of the concepts. The basic difficulty is a very practical one in that much of our experimental measurements in rock mechanics are essentially measurements of displacements.

Strains are calculated from displacements, while stresses are calculated from strains for comparison with theory. Naturally, it is desirable to simplify this sequence of measurement and inference whenever the conditions at hand permit.

Generally speaking, linearization that results from neglecting contributions of second- and higher-order terms is the main source of simplification. In an expression such as $1 + \varepsilon + \varepsilon^2 + O(\varepsilon^3)$, the term ε^2 and terms of higher order may be considered negligible compared with ε. The linear term ε^2 must then be "small" in this sense and considerably less than unity. For example, if $\varepsilon = 0.001$, then $\varepsilon^2 = 0.000\,001$, so $\varepsilon + \varepsilon^2 = 0.001\,001 \doteq 0.001$. Obviously, if ε is greater than 1, then ε^2 will certainly not be negligible with respect to ε. Care is needed when making such an approximation, and it should not be done too quickly. For example, $1 + \varepsilon^2 \doteq 1$, but $\sqrt{1 + \varepsilon^2} - 1 \doteq \varepsilon$ rather than 0 (expand in a Maclaurin series). Similarly, $1 + \varepsilon - 1 \doteq \varepsilon$ (not 0).

Consider the two-dimensional element shown in Figure 3.2. For convenience, the axes are re-oriented so that x is parallel to OA and y is parallel to OB. The x-direction normal strain for the material line OA according to (57) is

$$
\varepsilon_{xx} = \frac{O'A' - OA}{OA}
$$
$$
= \frac{\left[(\Delta x)^2 + (\Delta y_A)^2 \right]^{1/2} - \Delta a}{\Delta a}
$$
$$
= \frac{\left[(\Delta u_A + \Delta a)^2 + (\Delta v_A)^2 \right]^{1/2} - \Delta a}{\Delta a}
$$
$$
= \left[\left(1 + \frac{\Delta u_A}{\Delta a} \right)^2 + \left(\frac{\Delta v_A}{\Delta a} \right)^2 \right]^{1/2} - 1
$$
$$
\varepsilon_{xx} = \left[\left(1 + \frac{\partial u}{\partial a} \right)^2 + \left(\frac{\partial v}{\partial a} \right)^2 \right]^{1/2} - 1
$$

where the displacements are used to compute $\Delta x = \Delta a + u(A) - u(O)$, $\Delta y_a = v(A) - v(O)$, $\Delta u_A = u(A) - u(O)$, and $\Delta v_A = v(A) - v(O)$. The labels A and O represent the corresponding points in Figure 3.2, and the last step is obtained by passing to the limit of the difference quotients. A similar procedure for the y-direction leads to the expression

$$
\varepsilon_{yy} = \left[\left(1 + \frac{\partial v}{\partial b} \right)^2 + \left(\frac{\partial u}{\partial b} \right)^2 \right]^{1/2} - 1
$$

The shearing strain is somewhat more involved because what is desired according to the definition (58) is the sum of the angles $\alpha_1 + \alpha_2$ that corresponds to the change in orientation of OB relative to OA. To compute, one first calculates the angle between the line segments O'B' and O'A' after deformation and then uses the fact that $\gamma = \alpha_1 + \alpha_2 = \pi/2 - \theta$ as shown in Figure 3.2. The formula for the cosine of the angle between the two line segments is $\cos(\theta) = \cos(\alpha_1)\cos(\beta_2) + \cos(\beta_1)\cos(\alpha_2)$ in the two-dimensional case examined here. From Figure 3.2 and with $O'A' = OA(1 + \varepsilon_{xx})$, $OA = \Delta a$, $OB = \Delta b$, $\Delta x = \Delta a + \Delta u_A$, $\Delta y = \Delta b + \Delta v_A$, $\Delta x_B = \Delta u_B$, $\Delta y_B = \Delta v_A$, $\alpha_1 + \alpha_2 = \pi/2 - \theta$ and the expressions for the normal strains obtained previously, one obtains a succession of

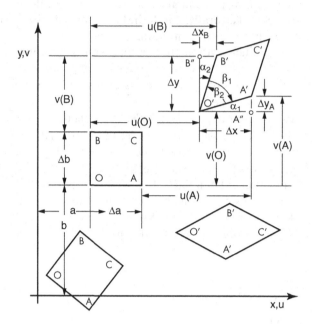

Figure 3.2 Geometric details in the geometry of primitive strain definitions.

steps leading to an expression from which the orientation change may be computed. Thus,

$$
\begin{aligned}
\cos(\theta) &= \left(\frac{O'A''}{O'A'}\right)\left(\frac{B'B''}{O'B'}\right) + \left(\frac{A'A''}{O'A'}\right)\left(\frac{O'B''}{O'B'}\right) \\
&= \left(\frac{\Delta x \Delta y_B + \Delta y_A \Delta y}{(O'A')(O'B')}\right) \\
&= \left(\frac{\Delta x \Delta y_B + \Delta y_A \Delta y}{(\Delta a \Delta b)(1+\varepsilon_{xx})(1+\varepsilon_{yy})}\right) \\
&= \left(\frac{\left(1+\dfrac{\Delta u_A}{\Delta a}\right)\left(\dfrac{\Delta u_B}{\Delta b}\right) + \left(1+\dfrac{\Delta v_B}{\Delta b}\right)\left(\dfrac{\Delta v_A}{\Delta a}\right)}{(1+\varepsilon_{xx})(1+\varepsilon_{yy})}\right)
\end{aligned}
$$

$$
\sin(\alpha_1 + \alpha_2) = \left(\frac{\left(1+\dfrac{\partial u}{\partial a}\right)\left(\dfrac{\partial u}{\partial b}\right) + \left(1+\dfrac{\partial v}{\partial b}\right)\left(\dfrac{\partial v}{\partial a}\right)}{\left[\left(1+\dfrac{\partial u}{\partial a}\right)^2 + \left(\dfrac{\partial v}{\partial a}\right)^2\right]^{1/2}\left[\left(1+\dfrac{\partial v}{\partial b}\right)^2 + \left(\dfrac{\partial u}{\partial b}\right)^2\right]^{1/2}}\right)
$$

The inverse sine gives the desired angle. Clearly, a straightforward application of the "elementary" definitions of strain leads to complicated nonlinear expressions for strains in terms of displacement derivatives.

However, if the displacement derivatives are small in the sense that products of derivatives are much smaller and may therefore be neglected, then great simplification occurs. With the assumption of small displacement derivatives, one has $\varepsilon_{xx} = \dfrac{\partial u}{\partial a}$, $\varepsilon_{yy} = \dfrac{\partial v}{\partial b}$, $\gamma_{xy} = \dfrac{\partial u}{\partial b} + \dfrac{\partial v}{\partial a}$, where the last follows from $\gamma = \alpha_1 + \alpha_2 = \gamma_{xy}$ and the fact that for small angles $\sin(\alpha_1 + \alpha_2) = \alpha_1 + \alpha_2$. These strains are "small" because of the approximation made. For example, $\varepsilon/(1+\varepsilon) = \varepsilon$ because $\varepsilon/(1+\varepsilon) = (\varepsilon - \varepsilon^2)/(1 - \varepsilon^2)$. This result is sometimes justified by saying that $(1+\varepsilon) = 1$, which is dangerous because $(1+\varepsilon) - 1 = \varepsilon$, not 0.

Another result of the "small" strain approximation is that the distinction between derivatives evaluated in the initial and final positions is of no consequence. Applying the chain rule for differentiation, one obtains

$$\begin{aligned}
\frac{\partial u}{\partial a} &= \frac{\partial u}{\partial x}\frac{\partial x}{\partial a} + \frac{\partial u}{\partial y}\frac{\partial y}{\partial a} \\
&= \frac{\partial u}{\partial x}\frac{\partial(a+u)}{\partial a} + \frac{\partial u}{\partial y}\frac{\partial y}{\partial a} \\
&= \frac{\partial u}{\partial x}\left(1 + \frac{\partial u}{\partial a}\right) + \frac{\partial u}{\partial y}\frac{\partial y}{\partial a} \\
\frac{\partial u}{\partial a} &= \frac{\partial u}{\partial x}
\end{aligned}$$

where $x = a + u$ and the last step is justified by neglecting products of derivatives in accordance with the small strain approximation. A similar analysis can be done for other derivatives, and indeed, the entire analysis can be carried out in three dimensions. The three-dimensional *small* strain–displacement equations that result are

$$\begin{aligned}
\varepsilon_{xx} &= \frac{\partial u}{\partial x}, & 2\varepsilon_{xy} = \gamma_{xy} &= \frac{\partial u}{\partial y} + \frac{\partial v}{\partial x} \\
\varepsilon_{yy} &= \frac{\partial v}{\partial y}, & 2\varepsilon_{yz} = \gamma_{yz} &= \frac{\partial v}{\partial z} + \frac{\partial w}{\partial y} \\
\varepsilon_{zz} &= \frac{\partial w}{\partial z}, & 2\varepsilon_{zx} = \gamma_{zx} &= \frac{\partial w}{\partial x} + \frac{\partial u}{\partial v}
\end{aligned} \tag{60}$$

where u, v, and w are the displacement components in the x-, y-, and z-directions, respectively. Examination of the geometry of the derivation leading to (60) shows $\varepsilon_{xy} = \varepsilon_{yx}$; that is, the shear strains are equal. The same results hold in three dimensions, so $\varepsilon_{yz} = \varepsilon_{zy}$ and $\varepsilon_{zx} = \varepsilon_{xz}$.

Thus, there is no distinction in magnitudes of the shear strain pairs, although they are physically associated with different surfaces.

The strains defined by (60) are used extensively, almost exclusively, in practical applications. The term "infinitesimal" strain is often used rather than "small" strain. Neither is entirely satisfactory because any strain is finite in the sense that it is not zero, and there can be relatively large infinitesimal strains and small "finite" strains as well. Usually, the context of a particular problem makes the meaning clear. In practice, one can always test numerically to see if ε^2 is negligible with respect to ε and use the result as a guide to deciding whether the assumption of small strains is reasonable. If not, then a different approach is indicated.

3.4   Geometric interpretation of small strains

The small strain–displacement relations have the same geometric interpretation originally used to define normal and shear strains. Figure 3.3 illustrates a geometric interpretation in more detail. The last case in Figure 3.3 that shows a rigid body rotation in the neighborhood of a point deserves some comment. Consider the components of the relative displacement of two nearby points. In two dimensions,

$$du = \frac{\partial u}{\partial x}dx + \frac{\partial u}{\partial y}dy$$

$$= \varepsilon_{xx}dx + \left(\frac{1}{2}\right)\left(\frac{\partial u}{\partial y} + \frac{\partial v}{\partial x}\right)dy + \left(\frac{1}{2}\right)\left(\frac{\partial u}{\partial y} - \frac{\partial v}{\partial x}\right)dy$$

$$du = \varepsilon_{xx}dx + \varepsilon_{yx}dy + \omega_{yx}dy$$

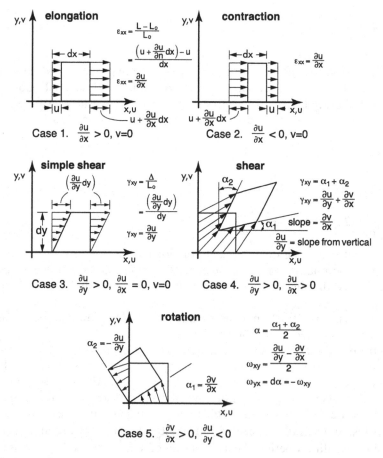

Figure 3.3 Geometric interpretation of "small" strain.

where ω_{yx} is an infinitesimal rotation of the considered element about the z-axis. A similar analysis holds in the y-direction. Thus,

$$dv = \frac{\partial v}{\partial y}dy + \frac{\partial v}{\partial x}dx$$

$$= \varepsilon_{yy}dx + \left(\frac{1}{2}\right)\left(\frac{\partial v}{\partial x} + \frac{\partial u}{\partial y}\right)dx + \left(\frac{1}{2}\right)\left(\frac{\partial v}{\partial x} - \frac{\partial u}{\partial y}\right)dy$$

$$du = \varepsilon_{yy}dx + \varepsilon_{xyx}dy + \omega_{xy}dy$$

where $\omega_{xy} = (1/2)(\partial v/\partial x - \partial u/\partial y) = -\omega_{yx}$; the rotations are anti-symmetric. Accordingly, if the strains vanish, then no deformation occurs, the element motion is rigid body, and the *relative* displacements describe a local rotation only. The body considered may translate, but translation produces no relative motion. Of course, if the motion is rigid body, then the converse is true and no deformation occurs.

To see that $\omega_{xy} = -\omega_{yx}$ is a "small" angle of rotation, consider the element shown in Figure 3.4a that experiences rigid body translation and rotation and then remove the translation with the result in Figure 3.4b, where the displacement components of P relative to O are

$$du = dr\cos(\beta_2) - dr\cos(\beta_1)$$
$$= dr[\cos(\beta_2)\cos(\alpha) - \sin(\beta_1)\sin(\alpha) - \cos(\beta_1)]$$
$$dv = dr[\sin(\beta_1)\cos(\alpha) - \sin(\beta_1)\sin(\alpha) - \sin(\beta_1)]$$

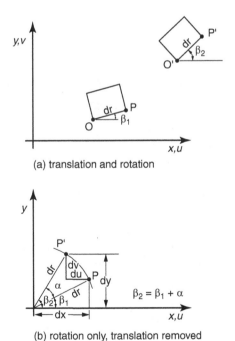

(a) translation and rotation

(b) rotation only, translation removed

Figure 3.4 Rigid body translation and rotation in the neighborhood of a point.

For small α,

$$du = -dr\sin(\beta_1)\alpha, \ dv = dr\cos(\beta_1)\alpha$$
$$du = -\alpha dy, \ dv = \alpha dx$$

Hence,

$$\alpha = \left(\frac{1}{2}\right)\left(\frac{\partial v}{\partial x} - \frac{\partial u}{\partial y}\right) = -\left(\frac{1}{2}\right)\left(\frac{\partial u}{\partial y} - \frac{\partial v}{\partial x}\right)$$

Alternatively, vanishing of the shear strains implies $\partial v/\partial x = -\partial u/\partial y$ and $(1/2)(\partial v/\partial x - \partial u/\partial y) = \partial v/\partial x = (1/2)(\alpha_1 + \alpha_2) = \alpha$. Thus, the relative displacement of neighboring points is composed of normal strain, shear strain, and a local rotation. Usually, the local rigid body motion is ignored because focus is on the strains, although occasionally the rotations are needed

3.5   Change of reference axes

In many practical problems, a change of reference axes simplifies the analysis. In the case of strains, the analysis proceeds in two steps. The displacements in the given system are first referred to the rotated axes, and then derivatives of these new displacements are computed in the rotated system. The results are then used according to the definition of strain. A two-dimensional analysis illustrates the procedure.

Consider the two coordinate systems shown in Figure 3.5, where the n, s, t system is rotated counterclockwise through an angle a with respect to the x, y, z system. Let

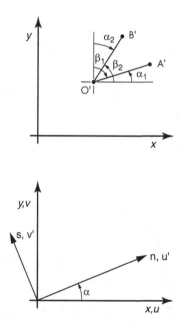

Figure 3.5 Axes rotation and displacements referred to the rotated axes.

the displacement components in the x- and y-directions be u and v, respectively, and the components in the n- and s-directions be u and v', respectively. The relationships between the two systems are

$$n = x\cos(\alpha) + y\sin(\alpha), \quad x = n\cos(\alpha) - s\sin(\alpha)$$
$$s = -x\sin(\alpha) + y\cos(\alpha), \quad y = n\sin(\alpha) + s\cos(\alpha)$$
and $\qquad\qquad\qquad\qquad\qquad\qquad\qquad\qquad\qquad\qquad$ (61)
$$u' = u\cos(\alpha) + v\sin(\alpha), \quad u = u'\cos(\alpha) - v'\sin(\alpha)$$
$$v' = -u\sin(\alpha) + v\cos(\alpha), \quad v = u'\sin(\alpha) + v'\cos(\alpha)$$

A shorthand notation is useful in the following: $C = \cos(\alpha)$, $S = \sin(\alpha)$. In the n, s system and after substitutions for du, dv, dx, and dy, one obtains

$$
\begin{aligned}
du' &= \frac{\partial u'}{\partial n}dn + \frac{\partial u'}{\partial s}ds \\
&= Cdu + Sdv \\
&= C\left(\frac{\partial u}{\partial x}dx + \frac{\partial u}{\partial y}dy\right) + S\left(\frac{\partial v}{\partial x}dx + \frac{\partial v}{\partial y}dy\right) \\
&= \left(C\frac{\partial u}{\partial x} + S\frac{\partial v}{\partial x}\right)(Cdn - Sds) + \left(C\frac{\partial u}{\partial y} + S\frac{\partial v}{\partial x}\right)(Sdn + Cds)
\end{aligned}
$$

$$
\begin{aligned}
du' &= \left(C^2\frac{\partial u}{\partial x} + SC\frac{\partial v}{\partial x} + CS\frac{\partial u}{\partial y} + S^2\frac{\partial v}{\partial x}\right)dn \\
&\quad + \left(-SC\frac{\partial u}{\partial x} - S^2\frac{\partial v}{\partial x} + C^2\frac{\partial u}{\partial y} + CS\frac{\partial v}{\partial x}\right)ds
\end{aligned}
$$

Comparison of the coefficients of dn shows that

$$\varepsilon_{nn} = C^2\varepsilon_{xx} + S^2\varepsilon_{yy} + 2SC\varepsilon_{xy}$$

where the small strain–displacement relations have been used. The coefficient of ds is

$$\frac{\partial u'}{\partial s} = -SC(\varepsilon_{xx} - \varepsilon_{yy}) - S^2\frac{\partial v}{\partial x} + C^2\frac{\partial u}{\partial y}$$

A similar analysis for v' shows that

$$\varepsilon_{ss} = S^2\varepsilon_{xx} + C^2\varepsilon_{yy} - 2SC\varepsilon_{xy}$$

and

$$\frac{\partial v'}{\partial n} = -SC(\varepsilon_{xx} - \varepsilon_{yy}) - S^2\frac{\partial u}{\partial y} + C^2\frac{\partial v}{\partial x}$$

Adding the derivatives and using the definition of shear strain gives

$$\varepsilon_{ns} = -SC(\varepsilon_{xx} - \varepsilon_{yy}) - (C^2 - S^2)\varepsilon_{xy}$$

Collecting results gives the two-dimensional part of the rotation of reference axes about the original z-axis. Thus,

$$\varepsilon_{nn} = (1/2)\,(\varepsilon_{xx} + \varepsilon_{yy}) + (1/2)\,(\varepsilon_{xx} - \varepsilon_{yy})\cos(2\alpha) + \varepsilon_{xy}\sin(2\alpha)$$
$$\varepsilon_{ss} = (1/2)\,(\varepsilon_{xx} + \varepsilon_{yy}) - (1/2)\,(\varepsilon_{xx} - \varepsilon_{yy})\cos(2\alpha) - \varepsilon_{xy}\sin(2\alpha) \qquad (62a)$$
$$\varepsilon_{ns} = \varepsilon_{sn} = (-1/2)\,(\varepsilon_{xx} - \varepsilon_{yy})\sin(2\alpha) + \varepsilon_{xy}\cos(2\alpha)$$

The z-direction strains transform according to

$$\varepsilon_{tt} = \varepsilon_{zz}$$
$$\varepsilon_{tn} = \varepsilon_{zx}\cos(\alpha) + \varepsilon_{zy}\sin(\alpha) \qquad (62b)$$
$$\varepsilon_{ts} = -\varepsilon_{zx}\sin(\alpha) + \varepsilon_{zy}\cos(\alpha)$$

These transformation Equations (62a and b) are identical in form to the two-dimensional equations of transformation of stress under a rotation of the reference axes (34a and b). The three-dimensional form for strains is also the same as that for stress (34c).

3.6   Principal strains, maximum shear strain, and Mohr's circle

Because the transformation of strain is mathematically identical to the transformation of stress, one can immediately arrive at expressions for principal strains, maximum and minimum shear strains, and so forth. The principal strains are given by

$$\left\{\begin{matrix}\varepsilon_1\\\varepsilon_3\end{matrix}\right\} = \left(\frac{1}{2}\right)(\varepsilon_{xx} + \varepsilon_{yy}) + \left[\left(\frac{(\varepsilon_{xx} - \varepsilon_{yy})}{2}\right)^2 + \left(\varepsilon_{xy}\right)\right]^{1/2} \qquad (63)$$

The angle from the x-axis to the direction of ε_1 may be obtained from

$$\tan(2\alpha) = \frac{\varepsilon_{xy}}{(1/2)\,(\varepsilon_{xx} - \varepsilon_{yy})} \qquad (64)$$

However, there is nothing in (64) that guarantees the solution to agree with the same angles obtained from (36); the directions of principal stress and strain do not necessarily coincide. In case the material of interest lacks any directional features and is therefore *isotropic*, then the principal directions coincide. If the material is *anisotropic*, a slate for example, then the principal directions may not coincide depending on the specifics of the problem at hand.

The maximum and minimum of (mathematical) shear strain are

$$\left\{\begin{matrix}\gamma_{max}\\\gamma_{min}\end{matrix}\right\} = \left[\left(\frac{(\varepsilon_{xx} - \varepsilon_{yy})}{2}\right)^2 + \left(\varepsilon_{xy}\right)\right]^{1/2} = \left(\frac{\varepsilon_1 - \varepsilon_3}{2}\right) \qquad (65)$$

and are associated with planes having normals inclined at angles to the x-axis that may be obtained from

$$\tan(2\alpha) = \frac{(1/2)\,(\varepsilon_{xx} - \varepsilon_{yy})}{\varepsilon_{xy}} \tag{66}$$

the solutions of which are at $\pm\pi/4$ to the direction of the major principal strain just as in the case of stress.

The strain equations of transformation have a graphical representation, a Mohr's circle, in a normal strain shear strain plot. The center of the circle is located on the normal stress axis at $(1/2)\,(\varepsilon_1 + \varepsilon_3)$; the circle radius is $(1/2)\,(\varepsilon_1 - \varepsilon_3)$, as illustrated in Figure 3.6, where compression is considered positive. Again, what was said about Mohr's circle for stress carries entirely over to Mohr's circle for strain including the possibility of representing a three-dimensional state of strain with three related circles.

3.7 Volumetric and deviatoric strains

A change in material volume measured per unit of original volume is the volumetric strain. If the original volume of a small generic cube is $\Delta a\Delta b\Delta c$, the product of the cube edge, and the volume after deformation is $\Delta x\Delta y\Delta z$, then the volumetric strain $\varepsilon_V = (\Delta V - \Delta V_o)/\Delta V_o$. The final lengths of the cube edge are $\Delta x = \Delta a + \varepsilon_{xx}\Delta a$, $\Delta y = \Delta b + \varepsilon_{yy}\Delta b$, $\Delta z = \Delta c + \varepsilon_{zz}\Delta c$, which leads to the expression

$$\frac{\Delta V - \Delta V_o}{\Delta V_o} = \frac{\Delta a\Delta b\Delta c\,(1+\varepsilon_{xx})\,(1+\varepsilon_{yy})\,(1+\varepsilon_{zz})}{\Delta a\Delta b\Delta c}$$

After expansion and introduction of the small strain approximation, the result is

$$\varepsilon_V = \varepsilon_{xx} + \varepsilon_{yy} + \varepsilon_{zz} \tag{67}$$

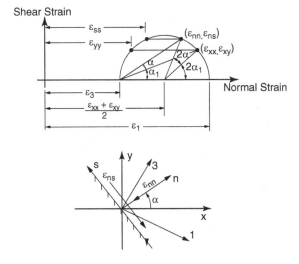

Figure 3.6 Mohr's circle for strain.

The volumetric strain given in (67) is also the first invariant of strain much as the sum of the normal stresses is the first invariant of stress in three dimensions.

Deviatoric strains are defined in the same manner as deviatoric stress. The deviatoric normal strains are then the normal strains less the mean normal strain, while the deviatoric shear strains are just the shear strains. Thus,

$$
\begin{aligned}
e_{xx} &= \varepsilon_{xx} - e, \; e_{xy} = \varepsilon_{xy} \\
e_{yy} &= \varepsilon_{yy} - e, \; e_{yz} = \varepsilon_{yz} \\
e_{zz} &= \varepsilon_{zz} - e, \; e_{zx} = \varepsilon_{zx}
\end{aligned}
\tag{68}
$$

where $e = (1/3)(\varepsilon_{xx} + \varepsilon_{yy} + \varepsilon_{zz}) = \varepsilon_V$. Deviatoric strains are generally associated with shape change. However, even if the shear strains vanish, shape changes may occur. For example, compression of a cube into a rectangular brick under the action of normal strains only, as shown in Figure 3.1, involves a change in shape as well as volume. The deviatoric part of normal strain is associated with shape change, while the volumetric part relates to the change in volume per unit of original volume.

3.8 Small strain compatibility

A one-to-one correspondence between material points before and after deformation is expected on physical grounds. Otherwise, gaps and overlaps of material would be indicated. This expectation implies that displacements are single-valued and continuous mathematical functions. The order of differentiation does not matter. Thus,

$$
\frac{\partial \varepsilon_{xx}}{\partial y} = \frac{\partial^2 u}{\partial y \partial x}, \; \frac{\partial \varepsilon_{yy}}{\partial x} = \frac{\partial^2 v}{\partial x \partial y}, \; 2\frac{\partial \varepsilon_{xy}}{\partial y} = \frac{\partial^2 u}{\partial y^2} + \frac{\partial^2 v}{\partial y \partial x}
$$

Differentiating again gives the relations

$$
\frac{\partial^2 \varepsilon_{xx}}{\partial y^2} = \frac{\partial^3 u}{\partial y^2 \partial x}, \; \frac{\partial^3 \varepsilon_{yy}}{\partial x^2} = \frac{\partial^2 v}{\partial x^2 \partial y}, \; 2\frac{\partial^2 \varepsilon_{xy}}{\partial y \partial x} = \frac{\partial^3 u}{\partial y^2 \partial x} + \frac{\partial^3 v}{\partial y \partial x^2}
$$

Hence,

$$
\frac{\partial^2 \varepsilon_{xx}}{\partial y^2} + \frac{\partial^3 \varepsilon_{yy}}{\partial x^2} = 2\frac{\partial^2 \varepsilon_{xy}}{\partial y \partial x}
\tag{69}
$$

which expresses the two-dimensional *compatibility* requirement for strains. If the displacements are single-valued, continuous, and differentiable, then (69) holds. However, if (69) does not hold, then the displacements are not single-valued. Additional but quite similar conditions can be obtained in three dimensions.

3.9 Strain and deformation rate

The time rate of change of strain for a particular material element is

$$
\frac{D\varepsilon(x,y,z,t)}{Dt} = \frac{\partial \varepsilon}{\partial t} + \frac{\partial \varepsilon}{\partial x}v_x + \frac{\partial \varepsilon}{\partial y}v_y + \frac{\partial \varepsilon}{\partial z}v_z
\tag{70}
$$

where the partial derivative is the rate of change at a fixed point and the remaining terms on the right are the convective part associated with element motion.

In using a superior dot to indicate a time derivative, care must be exercised to distinguish between the total derivative on the left of (70) and the partial derivative term on the right. Sometimes strain rates are given as

$$
\dot{\varepsilon}_{xx} = \frac{\partial v_x}{\partial x}, \quad 2\dot{\varepsilon}_{xy} = \dot{\gamma}_{xy} = \frac{\partial v_x}{\partial y} + \frac{\partial v_y}{\partial x}
$$
$$
\dot{\varepsilon}_{yy} = \frac{\partial v_y}{\partial y}, \quad 2\dot{\varepsilon}_{yz} = \dot{\gamma}_{yz} = \frac{\partial v_y}{\partial z} + \frac{\partial v_z}{\partial y} \tag{71}
$$
$$
\dot{\varepsilon}_{zz} = \frac{\partial v_z}{\partial z}, \quad 2\dot{\varepsilon}_{zx} = \dot{\gamma}_{zx} = \frac{\partial v_z}{\partial x} + \frac{\partial v_x}{\partial z}
$$

but these are more properly defined as rates of deformation. Fortunately, the distinction is not important when the strains are small and the convective part of (70) is negligible.

If one considers small changes in velocity, then the change can be expressed as a normal rate of deformation, a shear rate, and a rate of local rotation. Thus,

$$
dv_x = \dot{\varepsilon}_{xx}dx + \dot{\varepsilon}_{xy}dy + \dot{\omega}_{xy}dy
$$
$$
dv_y = \dot{\varepsilon}_{yy}dy + \dot{\varepsilon}_{yx}dx + \dot{\omega}_{yx}dx \tag{72}
$$

analogous to decomposition of relative displacement into normal and shear strain parts and a local rotation. The local rotation rate is

$$
-\dot{\omega}_{xy} = \dot{\omega}_{yx} = -\left(\frac{1}{2}\right)\left(\frac{\partial v_x}{\partial y} - \frac{\partial v_y}{\partial x}\right) = \frac{d\alpha}{dt} = \dot{\alpha}
$$

Again the dot time derivative, strictly speaking, is the total time derivative and that in the small strain approximation is a partial time derivative.

4   STRESS–STRAIN LAWS – ELASTICITY

Stress–strain laws are material laws. Unlike physical laws and kinematic relations that hold for all materials, stress–strain laws depend on the "constitution" of the material. Material laws are also called constitutive equations; they describe categories of material behavior. Some examples of material laws are: Hooke's law in elasticity, Ohm's law in electricity, Fourier's law in heat conduction, and Darcy's law for seepage in porous media. Although material laws are based on experimental observation, they are simplified models of actual material behavior.

One cannot avoid adoption of a material model in design analysis. In a strictly two-dimensional system, the conservation of mass ("continuity") equation may be expressed as

$$
\frac{\partial \rho}{\partial t} + \frac{\partial \rho v_x}{\partial x} + \frac{\partial \rho v_y}{\partial y} = 0 \tag{73}
$$

and the balance of linear momentum as

$$
\frac{\partial \sigma_{xx}}{\partial x} + \frac{\partial \tau_{xy}}{\partial y} + \gamma_x = \rho a_x
$$
$$
\frac{\partial \tau_{yx}}{\partial x} + \frac{\partial \sigma_{yy}}{\partial y} + \gamma_y = \rho a_y
$$

(74)

Equations (73) and (74) are a system of three equations containing six unknown functions of x and y (1 mass density, 3 stresses, and 2 velocities), where the specific weight γ is given as ρg (g is gravity acceleration) and accelerations are obtained as time rates of velocities. Velocities can be obtained from the displacements. Introduction of strains via the strain–displacement equations $\varepsilon_{xx} = \partial u / \partial x$ and so forth introduces three additional unknowns and also three additional equations. In three dimensions, the count is 10 equations (1 conservation of mass, 3 balance of linear momentum, and 6 strain–displacement equations) containing 16 unknowns (1 density, 6 stresses, 3 displacements, and 6 strains), where again velocities and accelerations are obtained by differentiating displacements. Hence, there is generally a deficit of 6 equations.

The presence of six stresses and six strains in the list of unknowns suggests that the additional equations needed may be obtained by viewing each of the stresses as a function of the strains. The opposite view could also be adopted, of course. Strains could be considered as functions of stress. Thus, when the stress–strain relations

$$
\sigma_{xx} = f_1(\varepsilon_{xx}, \varepsilon_{yy}, \varepsilon_{zz}, \gamma_{xy}, \gamma_{yz}, \gamma_{zx})
$$
$$
\sigma_{yy} = f_2(\varepsilon_{xx}, ..., \gamma_{zx})
$$
$$
\sigma_{zz} = f_3(\varepsilon_{xx}, ..., \gamma_{zx})
$$
$$
\tau_{xy} = f_4(\varepsilon_{xx}, ..., \gamma_{zx})
$$
$$
\tau_{yz} = f_5(\varepsilon_{xx}, ..., \gamma_{zx})
$$
$$
\tau_{zx} = f_6(\varepsilon_{xx}, ..., \gamma_{zx})
$$

are added to the equations derived from physical laws and kinematics, the deformation can be determined, at least, in principle. In practice, one would need to know the six functions that could be quite complex, especially if additional dependencies were present.

Two examples of added dependencies are temperature and strain rate. Temperature would require a seventh equation to be added to the list; strain rate as a derivative of strain would not increase the number of additional equations needed.

The form of the stress–strain relations may be postulated outright or determined experimentally. The first approach is based on experience with real materials, while the latter is guided by the postulated mathematical form. Clearly, the two approaches to specification of stress–strain relations are not completely independent. Hypotheses guide experimental inquiries, and experimental findings refine hypotheses until the accumulation of evidence justifies establishment of a stress–strain law. Hooke's law of linear elasticity of solids is an outstanding example of such a law.

The range of a purely elastic response of a solid to load is limited by strength, although some solids show a significant viscous contribution to deformation. Elasticity, strength, and viscosity lead to three classical and mathematically complex theories in the mechanics of solids – *elasticity*, *elastoplasticity* or simply *plasticity*, and *viscoelasticity*. Fortunately, a great many design problems of practical importance can be adequately addressed on the basis of a simplified theory of elasticity. The limits to the elastic range must still be known, so consideration of strength and failure criteria is essential, but the complexities of plasticity theory for geologic media may be avoided in elastic design. However, if the viscous response of the material is significant, then some account of the change of stress and strain with time is required, especially in design of excavations with relatively long service lives.

Still, the *de facto* standard model in solid mechanics, including rock mechanics, is elasticity. Elastic behavior in contrast to inelastic behavior implies reversible deformation. Work done during an elastic deformation is recoverable. Inelastic deformation is irreversible; not all work done during loading can be recovered as work upon unloading. This basic distinction makes Hooke's law and the limit to elastic deformation imposed by material strength of great importance to design analysis.

4.1 Hooke's law in one dimension – Young's modulus and shear modulus

The simplest non-trivial relationship between stress and strain is a linear one. Linearity and reversibility are the essence of Hooke's law. In the one-dimensional case, stress σ is directly proportional to strain ε. Although the real world is three-dimensional, a one-dimensional representation illustrates several important features of linear elasticity in a simple direct way. These features include elastic moduli, superposition, strain energy, stiffness, initial stress, and initial strain.

If the constant of proportionality between normal stress and normal strain is E, then Hooke's law in one dimension is simply

$$\sigma = E\varepsilon \tag{75}$$

A plot of (75) with ε as abscissa and σ as ordinate would obviously be a straight line passing through the origin, as shown in Figure 4.1a. The slope of the plot E is *Young's modulus*, an important elastic material property that retains in three dimensions the proportionality between normal strain and corresponding stress.

An important feature of linearity is *superposition*. If a given state of stress is $\sigma(1) = E\varepsilon(1)$ and a second state is $\sigma(2) = E\varepsilon(2)$, addition of strains $\varepsilon(1) + \varepsilon(2) = \varepsilon(3)$ implies addition of stresses such that $\sigma(1) + \sigma(2) = \sigma(3)$ because $\sigma(3) = E\varepsilon(1) + E\varepsilon(2)$. Superposition implies that states of stress and corresponding strain can simply be added to obtain the combined effects. Solutions to complicated problems may then be reduced to solution to several simpler problems. Linearity implies superposition, while nonlinearity precludes applicability of superposition.

In one dimension, the area under a stress–strain curve (line) represents the work done per unit volume of material during loading to the present state of strain. This work is stored in the body as *strain energy*. If A is the cross-sectional area of a prismatic bar and L is the length, then the average normal stress is $\sigma = F/A$ and the average normal strain is $\varepsilon = (L - L_0)/L_0$, as shown in Figure 4.1a. An element of area under the stress–strain curve $dw = \sigma d\varepsilon$, so

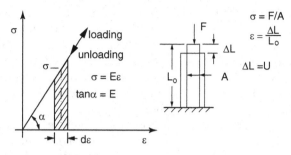

(a) one dimensional compression

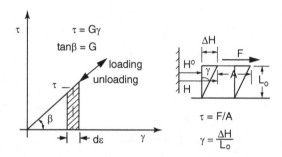

Figure 4.1 One-dimensional stress, strain, external work, and stored strain energy.

$$dw = \sigma\,d\varepsilon = \frac{R\,dL}{AL_o}$$

Hence,

$$w = \int_L \sigma\,d\varepsilon = \left(\frac{1}{AL_o}\right)\int_L F\,dL$$

where AL_o is the volume of the bar, the integral on the far right is the work done by the externally applied load, and w is the area under the stress–strain curve. In fact, w is the work per unit volume of the bar. Thus,

$$w = \int_L \sigma\,d\varepsilon = \int_L E\varepsilon\,d\varepsilon = \frac{E\varepsilon^2}{2} = \frac{\sigma\varepsilon}{2} = \frac{\sigma^2}{2E} \qquad (76)$$

expresses the external work in terms of the internal variables of stress and strain. When the work per unit volume done by the applied loads is expressed in terms of strain, the result is a *strain energy density*.

Young's modulus is closely associated with another elastic feature *normal stiffness*. However, stiffness is a structural feature and not an intrinsic material property. In an elastic structure composed of a number of connected elastic elements, an applied force is directly proportional to displacement; the constant of proportionality is a stiffness K_n. Thus, in one dimension, $F = K_n U$, where U is displacement in the direction of the applied force. If the structure is simply the prismatic bar considered previously, then the relevant displacement is the change in length of the sample, that is, $U = (L - L_o)$. Thus,

$$f = \sigma A = K_n U = K_n \varepsilon L_o + K_n \left(\frac{\sigma}{E} \right) L_o$$

Hence, the normal stiffness is

$$K_n = \left(\frac{AE}{L_o} \right) \tag{77}$$

Stiffness (normal stiffness) therefore depends not only on Young's modulus of the material, but also on the geometry of the structure, which is quite simple in the case of a prismatic bar, but less so in the case of a wood crib, for example. Where the geometry of a structure is more involved, the meaning of A and L_o is not always clear. Comparisons of stiffness of structures may also be unclear, but comparisons of Young's modulus of materials are unambiguous. In either case, comparisons must be made within the elastic range. Strength of material cannot be exceeded, nor can structural stability be violated. An important example of structural instability that may occur within the elastic range of material deformation is buckling of columns. The distinction between strength and stability is not merely semantics, but is an important physical distinction.

Application of load may be to a body that is already stressed or strained. Under such circumstances, changes in stress and strain that follow Hooke's law are of interest. Final states of stress and strain are obtained by adding the stress and strain changes to the initial values. For example, if a strain ε_o is present in the unstressed state as shown in Figure 4.2a, then the stress in one dimension is given by

$$\sigma = E(\varepsilon - \varepsilon_o)$$

Or if a stress σ_o is present in the unstrained state as shown in Figure 4.2b, then

$$(\sigma - \sigma_o) = E\varepsilon$$

Because of linearity of the stress–strain law, one can calculate a stress that corresponds to an initial strain and a strain that corresponds to an initial stress, as shown in Figure 4.2. Over a linear region shown in Figure 4.2c, one has

$$(\sigma - \sigma_o) = \Delta\sigma = E(\varepsilon - \varepsilon_o) = E\Delta\varepsilon$$

so increments of stress and strain are linearly related by Hooke's law in the elastic domain.

The incremental approach is often useful even in the nonlinear elastic case where Young's modulus E depends on strain, that is, where $\sigma = E(\varepsilon)\varepsilon$. If at a given strain, the change in E is small, then the stress change for practical purposes is linearly related to the strain change at the given strain. A common occurrence of these conditions is in experimental determination of Young's modulus by wave propagation through a test specimen under static load.

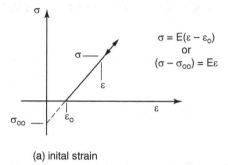

$$\sigma = E(\varepsilon - \varepsilon_0)$$
or
$$(\sigma - \sigma_{00}) = E\varepsilon$$

(a) inital strain

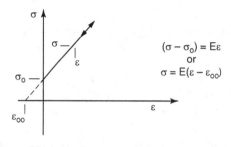

$$(\sigma - \sigma_0) = E\varepsilon$$
or
$$\sigma = E(\varepsilon - \varepsilon_{00})$$

(b) inital stress

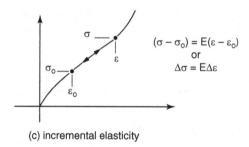

$$(\sigma - \sigma_0) = E(\varepsilon - \varepsilon_0)$$
or
$$\Delta\sigma = E\Delta\varepsilon$$

(c) incremental elasticity

Figure 4.2 Schematics of initial strain, stress, and incremental stress–strain relationships.

When the applied stress is a shear stress τ, then the corresponding strain is a shear strain γ (engineering shear strain). The constant of proportionality between the two is a *shear modulus*. Hooke's law in the case of one-dimensional shear is simply

$$\tau = G\gamma \tag{78}$$

where G is the shear modulus of the considered material. The shear modulus is also known as the modulus of rigidity and is often denoted by the symbol μ. Numerical values of G generally range between one-third and one half of E. A plot of τ as a function of γ is a straight line passing through the origin that has slope G, as shown in Figure 4.1b.

Shear strain energy density w is the work per unit volume done by the applied shear force acting through the shear displacement and is graphically represented by the area under the shear stress–shear strain curve. Thus,

$$w = \frac{G\gamma^2}{2} = \frac{\tau\gamma}{2} = \frac{\tau^2}{2G} \tag{79}$$

is the shear strain energy density. Shear stiffness K_s in contrast to shear modulus is

$$K_s = \frac{AG}{L_o} \tag{80}$$

as illustrated in Figure 4.1b. In structures such as wood cribs that have little resistance to side forces, shear stiffness may be only a small fraction of the wood shear modulus. Consideration of initial shear stress and strain and changes in shear stress and strain follows the same line of inquiry used for normal stress and strain.

4.2   Hooke's law in three dimensions – other elastic moduli

Application of normal stress, say, a tension σ, to a prismatic bar produces not only a normal strain in the direction of σ, but also a contraction across the bar. This effect is illustrated in Figure 4.3. Thus, even if the stress is one-dimensional, the strain is not. Additional strain terms are therefore needed in Hooke's law. Application of shear

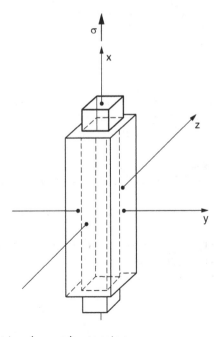

Figure 4.3 Poisson effect in a bar under tension.

stress may also change dimensions of the considered body and cause strains besides the associated shear strain. The possibility thus arises that stress depends on all six strains, that is,

$$\sigma = c_1 \varepsilon_{xx} + c_2 \varepsilon_{yy} + c_3 \varepsilon_{zz} + c_4 \gamma_{xy} + c_5 \gamma_{yz} + c_6 \gamma_{zx}$$

where the cs are elastic constants. When all six stresses are present, additional equations are

$$
\begin{aligned}
\sigma_{xx} &= b_{11}\varepsilon_{xx} + b_{12}\varepsilon_{yy} + b_{13}\varepsilon_{zz} + b_{14}\gamma_{xy} + b_{15}\gamma_{yz} + b_{16}\gamma_{zx} \\
\sigma_{yy} &= b_{21}\varepsilon_{xx} + b_{22}\varepsilon_{yy} + b_{23}\varepsilon_{zz} + b_{24}\gamma_{xy} + b_{25}\gamma_{yz} + b_{26}\gamma_{zx} \\
\sigma_{zz} &= b_{31}\varepsilon_{xx} + b_{32}\varepsilon_{yy} + b_{33}\varepsilon_{zz} + b_{34}\gamma_{xy} + b_{35}\gamma_{yz} + b_{36}\gamma_{zx} \\
\tau_{xy} &= b_{41}\varepsilon_{xx} + b_{42}\varepsilon_{yy} + b_{43}\varepsilon_{zz} + b_{44}\gamma_{xy} + b_{45}\gamma_{yz} + b_{46}\gamma_{zx} \\
\tau_{yz} &= b_{51}\varepsilon_{xx} + b_{52}\varepsilon_{yy} + b_{53}\varepsilon_{zz} + b_{54}\gamma_{xy} + b_{55}\gamma_{yz} + b_{56}\gamma_{zx} \\
\tau_{zx} &= b_{61}\varepsilon_{xx} + b_{62}\varepsilon_{yy} + b_{63}\varepsilon_{zz} + b_{64}\gamma_{xy} + b_{65}\gamma_{yz} + b_{66}\gamma_{zx}
\end{aligned}
\tag{81}
$$

where the coefficients of strain, the bs, are elastic constants. Although at first glance there appear to be 36 (6×6) elastic constants, the equality of shear stress and shear strain requires that $b_{12} = b_{21}$, $b_{13} = b_{31}$, and so on, to $b_{56} = b_{65}$, which reduces the number of elastic constants to 21 at most.

Equations (81) may be inverted to express strain as functions of stress. Thus,

$$
\begin{aligned}
\varepsilon_{xx} &= a_{11}\sigma_{xx} + a_{12}\sigma_{yy} + a_{13}\sigma_{zz} + a_{14}\tau_{xy} + a_{15}\tau_{yz} + a_{16}\tau_{zx} \\
\varepsilon_{yy} &= a_{21}\sigma_{xx} + a_{22}\sigma_{yy} + a_{23}\sigma_{zz} + a_{24}\tau_{xy} + a_{25}\tau_{yz} + a_{26}\tau_{zx} \\
\varepsilon_{zz} &= a_{31}\sigma_{xx} + a_{32}\sigma_{yy} + a_{33}\sigma_{zz} + a_{34}\tau_{xy} + a_{35}\tau_{yz} + a_{36}\tau_{zx} \\
\gamma_{xy} &= a_{41}\sigma_{xx} + a_{42}\sigma_{yy} + a_{43}\sigma_{zz} + a_{44}\tau_{xy} + a_{45}\tau_{yz} + a_{46}\tau_{zx} \\
\gamma_{yz} &= a_{51}\sigma_{xx} + a_{52}\sigma_{yy} + a_{53}\sigma_{zz} + a_{54}\tau_{xy} + a_{55}\tau_{yz} + a_{56}\tau_{zx} \\
\gamma_{zx} &= a_{61}\sigma_{xx} + a_{62}\sigma_{yy} + a_{63}\sigma_{zz} + a_{64}\tau_{xy} + a_{65}\tau_{yz} + a_{66}\tau_{zx}
\end{aligned}
\tag{82}
$$

Inspection of (82) shows that a stress σ in the x-direction produces a strain $\varepsilon_{xx} = a_{11}\sigma$, while the stress applied in the y-direction produces a strain $\varepsilon_{yy} = a_{22}\sigma$. Unless $a_{11} = a_{22}$, the strains will not be equal. If not, then the material has a directionally dependent elasticity; the material is *anisotropic*.

The most general anisotropy requires 21 elastic constants. If a material has three mutually orthogonal planes of symmetry and is therefore *orthotropic*, then nine elastic constants are needed. Gneisses are likely to be orthotropic. In case there is an axis of rotational symmetry, the material is *transversely isotropic* and requires five elastic constants. Sedimentary rock that has distinct bedding is likely to be transversely isotropic. Properties parallel and perpendicular to the bedding will differ, while axes orientation within a bedding plane is a matter of convenience.

In the absence of directionally dependent elasticity, the material is *isotropic*; properties are the same regardless of axes orientation. Massive sandstone and granite are likely to be isotropic. Only two independent elastic constants are needed to characterize an isotropic material. In the isotropic case, Hooke's law in the form (82) reduces to

$$\varepsilon_{xx} = \frac{1}{E}\sigma_{xx} - \frac{v}{E}\sigma_{yy} - \frac{v}{E}\sigma_{zz}, \quad \gamma_{xy} = \frac{1}{G}\tau_{xy}$$

$$\varepsilon_{yy} = \frac{-v}{E}\sigma_{xx} + \frac{1}{E}\sigma_{yy} - \frac{v}{E}\sigma_{zz}, \quad \gamma_{yz} = \frac{1}{G}\tau_{yz} \qquad (83)$$

$$\varepsilon_{zz} = \frac{-v}{E}\sigma_{xx} - \frac{v}{E}\sigma_{yy} + \frac{1}{E}\sigma_{zz}, \quad \gamma_{zx} = \frac{1}{G}\tau_{zx}$$

where v is *Poisson's ratio*, a new elastic constant. Because an isotropic material responds the same regardless of direction, rotation of the reference axis leaves (83) unchanged and leads to the relationships

$$v = \frac{E}{2G} - 1, \quad G = \frac{E}{2(1+v)}, \quad E = 2G(1+v) \qquad (84)$$

so that any one of the three elastic constants can be obtained from the other two. Under uniaxial stress as illustrated in Figure 4.3, the corresponding strain is $\varepsilon_a = \sigma_a/E$. The strain across the bar, the transverse strain, $\varepsilon_t = -v\sigma_a/E$. Hence,

$$v = \left|\frac{\varepsilon_t}{\varepsilon_a}\right| \qquad (85)$$

The transverse strain induced by load at right angles is often referred to as Poisson's ratio effect. Under tension, the transverse strain is a contraction, but under compression, a thickening occurs. Poisson's ratio varies between extremes of 0 and 0.5, although negative values are possible. As a practical matter, values of about 0.15–0.35 are common.

Another useful elastic constant is defined in terms of a volumetric strain–mean normal stress relation. Adding the first three of (83) to obtain the volumetric strain gives

$$(\varepsilon_{xx} + \varepsilon_{yy} + \varepsilon_{zz}) = \varepsilon_v = (\sigma_{xx} + \sigma_{yy} + \sigma_{zz})\left(\frac{1-2v}{E}\right) = 3\sigma_m\left(\frac{1-2v}{E}\right) = \frac{\sigma_m}{K} \qquad (86)$$

where ε_V is the volumetric strain, σ_m is the mean normal stress, and K is the *bulk modulus* of the material, $K = E/3 (1-2v)$. Poisson's ratio of 0.5 implies an incompressible material. Compressibility is the reciprocal of the bulk modulus and is zero when the material is incompressible.

Equations (83) in inverted form are

$$\sigma_{xx} = \left[\frac{E}{(1+v)(1-2v)}\right]\left[(1-v)\varepsilon_{xx} + v\varepsilon_{yy} + v\varepsilon_{zz}\right], \quad \tau_{xy} = G\gamma_{xy}$$

$$\sigma_{yy} = \left[\frac{E}{(1+v)(1-2v)}\right]\left[v\varepsilon_{xx} + (1-v)\varepsilon_{yy} + v\varepsilon_{zz}\right], \quad \tau_{yz} = G\gamma_{yz} \qquad (87)$$

$$\sigma_{zz} = \left[\frac{E}{(1+v)(1-2v)}\right]\left[v\varepsilon_{xx} + v\varepsilon_{yy} + (1-v)\varepsilon_{zz}\right], \quad \tau_{zx} = G\gamma_{zx}$$

An interesting feature of isotropic materials is the uncoupling of normal and shear components of stress and strain. Inspection of (83) and (87) shows that normal stresses are linked only to normal strains and that shear stresses are linked only to shear strain. This feature also holds for transversely isotropic and orthotropic materials when the reference axes coincide with the material axes.

Other useful forms of Hooke's law for isotropic materials can be obtained using volumetric strain ε_V. Thus,

$$
\begin{aligned}
\sigma_{xx} &= \lambda\varepsilon_v + 2G\varepsilon_{xx}, \quad \tau_{xy} = G\gamma_{xy} \\
\sigma_{yy} &= \lambda\varepsilon_v + 2G\varepsilon_{yy}, \quad \tau_{yz} = G\gamma_{yz} \\
\sigma_{zz} &= \lambda\varepsilon_v + 2G\varepsilon_{zz}, \quad \tau_{zx} = G\gamma_{zx}
\end{aligned}
\tag{88}
$$

where λ is a new elastic constant, one of two *Lamé* constants. The second *Lamé* constant is the shear modulus G. One has the relations

$$
\lambda = \frac{vE}{(1+v)(1-2v)} = K - \frac{2}{3}G, \quad K = \frac{3\lambda + 2G}{3}
\tag{89}
$$

Inspection of (89) shows that if $v = 0.25$, then $\lambda = G$ and $K = (5/3)G$.

$$
\begin{aligned}
\sigma_{xx} &= K\varepsilon_v + 2Ge_{xx}, \quad \tau_{xy} = G\gamma_{xy} \\
\sigma_{yy} &= k\varepsilon_v + 2Ge_{yy}, \quad \tau_{yz} = G\gamma_{yz} \\
\sigma_{zz} &= K\varepsilon_v + 2Ge_{zz}, \quad \tau_{zx} = G\gamma_{zx}
\end{aligned}
\tag{90}
$$

where e is deviatoric strain, for example $e_{xx} = \varepsilon_{xx} - \varepsilon_V/3$, by definition. Equations (90) show the contributions of volume and shape changes to normal stress. A number of relations among the various elastic constants are given in Table 4.1.

The work done by externally applied forces during deformation that is stored in a body as strain energy is given by

$$
\begin{aligned}
w = \int_V dw = \int_V \big(&\sigma_{xx}d\varepsilon_{xx} + \sigma_{yy}\,d\varepsilon_{yy} + \sigma_{zz}\,d\varepsilon_{zz} \\
&+ \tau_{xy}d\gamma_{xy} + \tau_{yz}d\gamma_{yz} + \tau_{zx}d\gamma_{zx}\big)
\end{aligned}
$$

Table 4.1 Elastic constants

Young's modulus	Shear modulus	Poisson's ratio	Bulk modulus	Lamé constant	Lamé constant*
E	G	v	K	λ	μ
$2G(1+v)$	$\dfrac{E}{2(1+v)}$	$\dfrac{(E-2G)}{2G}$	$\dfrac{EG}{3(3G-E)}$	$\dfrac{G(2G-E)}{(E-3G)}$	$\dfrac{E}{2(1+v)}$
$3K(1-2v)$	$\dfrac{3EK}{(9K-E)}$	$\dfrac{(3K-2G)}{2(3K+G)}$	$\dfrac{E}{3(1-2v)}$	$\dfrac{2Gv}{(1-2v)}$	$\dfrac{3EK}{(9K-E)}$
$\dfrac{\mu(3\lambda+2\mu)}{(\lambda+\mu)}$	$\dfrac{3K(1-2v)}{E(1+v)}$	$\dfrac{(3K-E)}{6K}$	$\dfrac{2G(1+v)}{3(1-2v)}$	$\dfrac{3Kv}{(1+v)}$	$\dfrac{3K(1-2v)}{E(1+v)}$
$\dfrac{9K\mu}{(3K+\mu)}$	$\dfrac{(3K-\lambda)}{2}$	$\dfrac{\lambda}{(3K-\lambda)}$	$\dfrac{3\lambda+2\mu}{3}$	$\dfrac{Ev}{(1+v)(1-2v)}$	$\dfrac{(3K-\lambda)}{2}$

* This constant μ is also the shear modulus G.

Use of Hooke's law allows for integration. Thus, the three-dimensional strain energy density as a quadratic form in strain is

$$w = \frac{K(\varepsilon_V)^2}{2} + G\left[\varepsilon_{xx}^2 + \varepsilon_{yy}^2 + \varepsilon_{zz}^2 + \left(\frac{1}{2}\right)\left(\gamma_{xy}^2 + \gamma_{yz}^2 + \gamma_{zx}^2\right)\right]$$

(91)

which shows volumetric and deviatoric strain contributions to the total strain energy density (strain energy per unit volume). Equation (91) can also be expressed in terms of mean normal stress and deviatoric stress, that is,

$$w = \frac{(\sigma_m)^2}{2K} + \left(\frac{1}{2G}\right)\left[\tau_{xy}^2 + \tau_{yz}^2 + \tau_{zx}^2 + \left(\frac{1}{2}\right)\left(s_{xx}^2 + s_{yy}^2 + s_{zz}^2\right)\right]$$

(92)

where s is the deviatoric stress, for example $s_{xx} = \sigma_{xx} - \sigma_m$, where σ_m is the mean normal stress. Equation (92) shows the contribution of the mean normal stress (hydrostatic part of stress) and the deviatoric stresses to the strain energy density. The bilinear form of the strain energy density is

$$w = \left(\frac{1}{2}\right)\left[\sigma_{xx}\varepsilon_{xx} + \sigma_{yy}\varepsilon_{yy} + \sigma_{zz}\varepsilon_{zz} + \tau_{xy}\gamma_{xy} + \tau_{yz}\gamma_{yz} + \tau_{zx}\gamma_{zx}\right]$$

which remains true when the material is anisotropic. An important theoretical property of the strain energy density function is that

$$\frac{\partial w}{\partial \varepsilon_{xx}} = \sigma_{xx}, \ldots, \quad \frac{\partial w}{\partial \gamma_{xy}} = \tau_{xy}, \ldots$$

$$\frac{\partial w}{\partial \sigma_{xx}} = \varepsilon_{xx}, \ldots, \quad \frac{\partial w}{\partial \tau_{xy}} = \gamma_{xy}, \ldots$$

which shows that the derivative of the strain energy density function with respect to strain gives the corresponding stress and vice versa.

4.3   More on elastic anisotropy

Any rock that has a preferred orientation of some of its constituent minerals either physically or crystallographically is likely to show some degree of anisotropy. Joints in large rock masses are often preferentially oriented and thus impart a direction character to the rock mass even when the intact rock between the joints is isotropic. Anisotropy is more the rule than the exception in rock mechanics. The question for the design analyst is therefore whether the degree of anisotropy is sufficient to warrant consideration in an analysis. Anisotropic behavior is far more complicated than isotropic behavior. However, numerical differences between anisotropic and isotropic analyses are often obscured by uncertainties in available design data, so unless the anisotropy is quite pronounced, the effects are often neglected within the elastic range of response.

Although elastic anisotropy may not be important in many practical situations, variation of strength with direction may be most significant. For example, Young's modulus measured parallel to the bedding of a laminated sedimentary rock may be as much as 50% greater than Young's modulus measured perpendicular to the bedding. Strengths measured parallel and perpendicular to the bedding may differ by several 100%. For this reason, strengths should be measured in the direction of anticipated loading, even though the analysis of stress may be based on the assumption of isotropic elastic behavior. More generally, strength anisotropy should be taken into account as indicated by geological evidence both at the laboratory scale of observation and at the scale of design and excavation.

When not negligible, anisotropy is usually treated according to two special cases, the orthotropic case or the transversely isotropic case. In the orthotropic case, the material has three mutually orthogonal planes of elastic symmetry. Intersections of planes of symmetry define three orthogonal material axes. Gneissic rock that has flattened mafics elongated in one direction is almost certain to be orthotropic. Rhyolite with flow structure may be orthotropic. Schists, phyllites, shales, and laminated sedimentary rocks are likely to show significant differences between properties parallel and perpendicular to the schistosity or bedding and thus to be transversely isotropic with no preferential direction about an axis perpendicular to the plane of stratification.

In the orthotropic case when one uses the material axes a, b, and c for reference, the generalized Hooke's law (81 and 82) reduces to

$$\varepsilon_{aa} = \frac{1}{E_a}\sigma_{aa} - \frac{\nu_{ba}}{E_b}\sigma_{bb} - \frac{\nu_{ca}}{E_c}\sigma_{cc}, \quad \gamma_{ab} = \frac{1}{G_c}\tau_{ab}$$

$$\varepsilon_{bb} = \frac{-\nu_{ab}}{E_a}\sigma_{aa} + \frac{1}{E_b}\sigma_{bb} - \frac{\nu_{cb}}{E_c}\sigma_{cc}, \quad \gamma_{bc} = \frac{1}{G}\tau_{bc} \tag{93}$$

$$\varepsilon_{cc} = \frac{-\nu_{ac}}{E_a}\sigma_{aa} - \frac{\nu_{bc}}{E_b}\sigma_{bb} + \frac{1}{E_c}\sigma_{cc}, \quad \gamma_{ca} = \frac{1}{G}\tau_{ca}$$

Although there appear to be 12 elastic constants in (93), 3 Es, 3 Gs, and 6 νs, symmetry reduces the number to nine ($E_a, E_b, E_c, G_a, G_a, G_a, \nu_{ab}, \nu_{bc}, \nu_{ca}$). The order of the subscripts with the Poisson's ratios is important. For example, ν_{ab} relates the effect of a stress in the a-direction to a strain in the b-direction, while ν_{ba} relates a stress in the b-direction to a strain in the a-direction. The two are generally not equal. However, symmetry requires

$$\frac{\nu_{ba}}{E_b} = \frac{\nu_{ab}}{E_a}, \quad \frac{\nu_{ca}}{E_c} = \frac{\nu_{ac}}{E_a}, \quad \frac{\nu_{cb}}{E_c} = \frac{\nu_{bc}}{E_b}, \tag{94}$$

which allows for computation of the other Poisson's ratios. Modulus E_a relates a stress applied in the a-direction to a strain in the same direction and similarly for E_b and E_c. A shear modulus G_c relates shear stress to shear strain in the ab-plane and similarly for G_a and G_b for shear in the bc- and ca-planes.

In the transversely isotropic case, one may consider bedding to be horizontal and the vertical axis to be the axis of rotational symmetry, as shown in Figure 4.4. Because transverse isotropy implies no distinction of direction in the horizontal plane, the material axes are simply vertical and horizontal, v and h. In this case, Hooke's law has the form

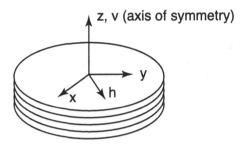

(a) transversely isotropic case

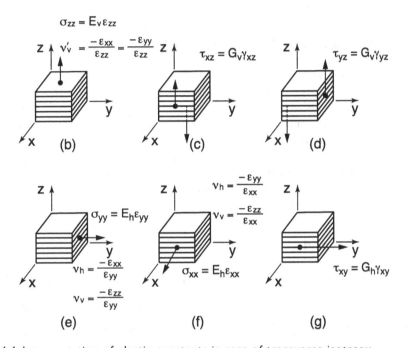

Figure 4.4 Interpretation of elastic constants in case of transverse isotropy.

$$\varepsilon_{xx} = \frac{1}{E_h}\sigma_{xx} - \frac{v_h}{E_h}\sigma_{yy} - \frac{v_v}{E_h}\sigma_{zz}, \quad \gamma_{xy} = \frac{1}{G_h}\tau_x$$

$$\varepsilon_{yy} = \frac{-v_h}{E_h}\sigma_{xx} + \frac{1}{E_h}\sigma_{yy} - \frac{v_v}{E_h}\sigma_{zz}, \quad \gamma_{yz} = \frac{1}{G_v}\tau_{yz} \qquad (95)$$

$$\varepsilon_{zz} = \frac{-v_v}{E_h}\sigma_{aa} - \frac{v_v}{E_h}\sigma_{yy} + \frac{1}{E_v}\sigma_{zz}, \quad \gamma_{zx} = \frac{1}{G_v}\tau_{zx}$$

The physical meanings of the various elastic constants in (95) are illustrated in Figure 4.4. Although there are six elastic constants in (95), 2 Es, 2 Gs, and 2 vs, only five are independent. In fact, the in-plane properties with subscript h are related by the isotropic formula

$$G_h = \frac{E_h}{2(1+v_h)} \tag{96}$$

The Poisson's ratio v_h relates a horizontal strain at right angles to a horizontal stress; v_v relates a vertical strain to a horizontal stress. However, v_v is not the same as v_v', which relates a horizontal strain to a vertical stress seen in Figure 4.4, although symmetry requires

$$\frac{v_v}{E_h} = \frac{v_v'}{E_v} \tag{97}$$

Moduli E_h and E_v relate horizontal and vertical stresses to corresponding strains; G_h and G_v, respectively, relate shear stress and strain parallel and perpendicular to the bedding, as illustrated in Figure 4.4.

5 PLANE STRESS, PLANE STRAIN, AND AXIAL SYMMETRY

There are three important special cases of three-dimensional stress and strain that reduce to mainly two-dimensional considerations. The first is *plane stress*; the second is *plane strain*; and the third is *axial symmetry*. The term two-dimensional implies that variation in a third direction is negligible. In plane stress and strain, this direction is the z-direction; in axial symmetry, variation with respect to the angular coordinate is nil. Analysis of the role of rock weight in estimating preexcavation stress, strain, and displacement is a practical plane problem. Calculation of stress, strain, and displacement changes induced by excavation of a vertical circular shaft in an elastic ground is an important application of axial symmetry.

5.1 Plane stress and plane strain

Plane stress and plane strain analyses are more complex than one-dimensional analyses, but not as complex as the general three-dimensional case. Both are illustrated in Figure 5.1. Many important practical problems are well approximated as plane stress or plane strain problems. Slabs and plates loaded on edge are examples of the former; tunnels and shafts are examples of the latter.

In *plane stress*, the three z-direction stresses are zero, while the three in-plane stresses remain to be determined. Thus, $\sigma_{zz} = \tau_{zy} = \tau_{zx} = 0$ in plane stress, while $\sigma_{xx}, \sigma_{yy}, \tau_{xy}$ are unknowns. Vanishing of the z-direction shear stresses implies through Hooke's law vanishing of the z-direction shear strains, so $\gamma_{zy} = \gamma_{zx} = 0$ and $E\varepsilon_{zz} = -v(\sigma_{xx} + \sigma_{yy})$. Solution to the two-dimensional problem for the stresses (assuming they are uniform through the thickness of the slab) then allows for computing ε_{zz} and the in-plane strains $\varepsilon_{xx}, \varepsilon_{yy}, \gamma_{xy}$. Hooke's law reduces to

$$\begin{aligned}
\varepsilon_{xx} &= \frac{1}{E}\sigma_{xx} - \frac{v}{E}\sigma_{yy} \\
\varepsilon_{yy} &= \frac{-v}{E}\sigma_{xx} + \frac{1}{E}\sigma_{yy} \\
\gamma_{xy} &= \frac{1}{G}\tau_{xy}
\end{aligned} \tag{98}$$

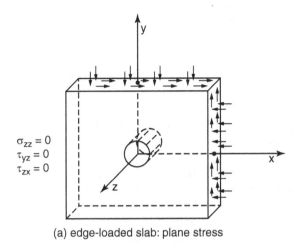

$$\sigma_{zz} = 0$$
$$\tau_{yz} = 0$$
$$\tau_{zx} = 0$$

(a) edge-loaded slab: plane stress

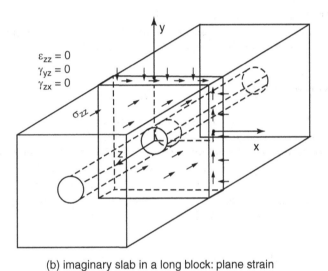

$$\varepsilon_{zz} = 0$$
$$\gamma_{yz} = 0$$
$$\gamma_{zx} = 0$$

(b) imaginary slab in a long block: plane strain

Figure 5.1 Concepts of plane stress (a) and plane strain (b).

in plane stress. Excavation of a hole in an edge-loaded slab induces stress and strain changes that follow the plane stress idealization. Interestingly enough, the change in slab thickness varies from point to point and is largest at the hole rim where a "lip" forms.

Plane strain may be defined by setting the z-direction strains to zero, so $\varepsilon_{zz} = \varepsilon_{zy} = \varepsilon_{zx} = 0$. The z-direction shear strains imply through Hooke's law that the z-direction shear stresses are zero (in the isotropic case). Also, the z-direction normal strain implies that $\sigma_{zz} = -v(\sigma_{xx} + \sigma_{yy})$. Again, solution for the in-plane stresses allows for determination of all stresses and strains. In the case of plane strain, Hooke's law has the form

$$\varepsilon_{xx} = \left(\frac{1-v^2}{E}\right)\sigma_{xx} - \left(\frac{v(1+v)}{E}\right)\sigma_{yy}$$

$$\varepsilon_{yy} = \frac{v(1+v)}{E}\sigma_{xx} + \left(\frac{1-v^2}{E}\right)\sigma_{yy} \qquad (99)$$

$$\gamma_{xy} = \frac{1}{G}\tau_{xy}$$

A thin imaginary slab perpendicular to the axis of a long tunnel located in a region where the pretunnel stress field shows no shear stress parallel to the tunnel axis is shown in Figure 5.1b. Excavation of the tunnel is reasonably idealized as a plane strain problem; that is, the changes in stress and strain induced by excavation satisfy plane strain conditions. Stresses and strains about the tunnel after excavation are obtained by addition of the changes to the pretunnel stresses and strains. Displacements induced by excavation are associated with the strain changes and are entirely in plane. Thickness of the considered slab remains unchanged in a plane strain analysis.

In anisotropic rock where the tunnel axis is not parallel to a material axis, vanishing of shear strains (stresses) no longer implies vanishing of the corresponding shear stresses (strains). Additional considerations are then required for plane stress and plane strain analyses that involve subtle details of the problem.

5.2 Rock weight and gravity stress

Considerations of stress and strain assist in determination of gravity-induced stresses, strains, and displacements that occur before any excavation. Although very often the specific weight of material in the vicinity of an excavation may be neglected in an analysis, why this is so is worth examining as is the case when weight is of interest. Equilibrium is required in any case. If the z-direction is vertical and the only vertical load acting on the considered rock mass is rock weight, then the z-direction shear stresses vanish and the stress equations of equilibrium reduce to

$$\frac{\partial \sigma_{xx}}{\partial x} + \frac{\partial \tau_{xy}}{\partial y} + \gamma_x = 0$$

$$\frac{\partial \tau_{yx}}{\partial x} + \frac{\partial \sigma_{yy}}{\partial y} + \gamma_y = 0 \qquad (100)$$

$$\frac{\partial \sigma_{zz}}{\partial z} + \gamma_z = 0$$

which include the special cases of plane stress and plane strain. In fact, the first two equations in (100) are identical to the strictly two-dimensional stress equations of equilibrium. With the vertical axis parallel to z, horizontal components of specific weight vanish, that is $\gamma_x = \gamma_y = 0$. However, the z-direction cannot be entirely ignored; the equilibrium requirement posed by the third equation in (100) must be met. At this juncture, neither the vertical stress nor the vertical strain is assumed to be zero. In case of plane stress, $\sigma_{zz} = 0$ (constant) and, of necessity, $\gamma_z = 0$, so weight of the slab cannot be of importance to the analysis. If the slab were thick and not really a slab, the weight could be important, but then a plane stress analysis would not be appropriate.

Application of weight also precludes plane strain because displacement in the z-direction γ_z is not zero and the action of "settlement" surely occurs under the action of gravity. Indeed, if γ_z is not zero, then the action of gravity is to be taken into account; from integration of the third equation in (100), $\sigma_{zz} = -\gamma z + f(x,y)$, where the subscript has been dropped from γ_Z. With the origin of coordinates at the surface of the earth, one has $\sigma_{zz} = 0 = f(x,y)$ for all points at the surface. Hence,

$$\sigma_{zz} = \gamma h \tag{101}$$

where h is the depth and compression is positive. A rule of thumb based on (101) is that the vertical stress, also known as the overburden stress or lithostatic stress, is 1 pound per square inch per foot of depth (1 psi/ft or 22.6 kPa/m). This rule is based on a specific weight of 144 pounds per cubic foot. The result (101) is a common assumption in rock mechanics and implies vanishing of the z-direction (vertical) shear stresses, so the vertical direction is a principal stress direction and horizontal planes are principal planes that are shear-free. Whether σ_{zz} is the major, minor, or intermediate principal stress is not known generally, but in the absence of evidence to the contrary and under gravity loading only, it is the major principal stress (compression positive). The horizontal stresses are often considered equal and are equal to some fraction or multiple of the vertical stress. Thus, with $\sigma_{xx} = \sigma_{yy} = \sigma_h$ and $\sigma_{zz} = \sigma_v$, one has

$$\sigma_h = K_o \sigma_v \tag{102}$$

where K_o is, by definition, the ratio of horizontal stress to vertical stress. In practice, K_o may be measured or estimated. In the absence of measurements and evidence for high horizontal stresses associated with forces of tectonic origin, σ_h is due to gravity alone. The considered region can then be viewed as a column of rock loaded by self-weight with the vertical stress given by (101), where γ is the average specific weight of the column. Three special cases are of interest:

1. a case where the block is free to expand laterally,
2. a case where no lateral expansion occurs, and
3. a case where the block cannot sustain a stress difference indefinitely.

In the last case, the block material deforms in a fluid-like manner and relieves any stress difference originally present. The first case is one of uniaxial stress because free lateral expansion implies no horizontal reaction. Hence, in the first case, $K_o = 0$. The second case is one of uniaxial (vertical) strain because no displacement is allowed in any horizontal direction. Hooke's law then shows that the horizontal stress is related to the vertical stress by

$$\sigma_h = \left(\frac{v}{1-v}\right)\sigma_v \tag{103}$$

so $K_o = v/(1-v)$. If $v = 0.25$, then $K = (1/3)$. In this case of complete lateral restraint, K ranges between 0 and 1 as Poisson's ratio ranges between the extremes of 0.0 and 0.5. Although the assumption of complete lateral restraint allows for an estimate of K_o from Poisson's ratio, the reverse is not possible. Generally, Poisson's ratio cannot be inferred from K_o. The third case of hydrostatic stress (and strain) implies $K_o = 1$. This case often occurs in salt formations. Flow of salt occurs slowly in geologic time, but eventually reaches equilibrium as the horizontal stresses become equal to the vertical stress. In case of "high" horizontal stress, K_o is greater than one; K_o could even be negative, although this case would be quite unusual.

If the earth is idealized as a half-plane that corresponds to the case of complete lateral restraint, then the only non-zero strain is in the vertical direction. According to Hooke's law and the strain–displacement relations

$$\varepsilon_{zz} = \frac{1}{E}\sigma_{zz} - \frac{v}{E}\sigma_{yy} - \frac{v}{E}\sigma_{xxy}$$

$$\frac{\partial w}{\partial z} = \left(\frac{1}{E}[1 - 2v\frac{v}{1-v}]\right)(\gamma z)$$

where z is positive down and tension is positive. Displacement is also positive down. After

$$w = \int (\text{constant})\, dz = (\text{constant})\left(\frac{z^2}{2}\right) + (\text{integration constant})$$

$$w = \left(\frac{\gamma}{2E}\right)\left[1 - 2v\left(\frac{v}{1-v}\right)(h^2 - z^2)\right]$$

(104)

where w is the vertical displacement of a point at depth z below the ground surface relative to a datum taken arbitrarily at a depth h. Alternatively, one may consider (104) as the vertical displacement in a layer of rock of thickness h, specific weight y, Young's modulus E, and Poisson's ratio v. If gravity were switched off, such a layer would rebound by the same amount. Removal of material during surface excavation effectively eliminates weight (nullifies gravity) and also induces rebound in the adjacent rock mass.

If the analysis of rock weight effect on preexcavation stress is carried out in cylindrical coordinates r, θ, z instead of Cartesian ("compass") coordinates x, y, z, then from the equations of transformation of reference axis

$$\sigma_{rr} = (1/2)(\sigma_{xx} + \sigma_{yy}) + (1/2)(\sigma_{xx} - \sigma_{yy})\cos(2\theta) + \tau_{xy}\sin(2\theta)$$
$$\sigma_{\theta\theta} = (1/2)(\sigma_{xx} + \sigma_{yy}) - (1/2)(\sigma_{xx} - \sigma_{yy})\cos(2\theta) + \tau_{xy}\sin(2\theta)$$
$$\tau_{r\theta} = (-1/2)(\sigma_{xx} - \sigma_{yy})\sin(2\theta) + \tau_{xy}\cos(2\theta)$$
$$\sigma_{zz} = \sigma_{zz}$$
$$\tau_{zr} = \tau_{zx}\cos(2\theta) + \tau_{zy}\sin(2\theta)$$
$$\tau_{z\theta} = -\tau_{zx}\sin(2\theta) + \tau_{zy}\cos(2\theta)$$

Hence,

$$\sigma_{rr} = \sigma_h$$
$$\sigma_{\theta\theta} = \sigma_h$$
$$\tau_{r\theta} = 0$$
$$\sigma_{zz} = \sigma_v$$
$$\tau_{zr} = 0$$
$$\tau_{z\theta} = 0$$

where $\sigma_{xx} = \sigma_{yy} = \sigma_h$, $\sigma_{zz} = \sigma_v$ and $\tau_{zx} = \tau_{zy} = \tau_{xy}$ under the assumption of gravity loading only. According to Hooke's law with respect to r, θ, z,

$$E\varepsilon_{rr} = \sigma_{rr} - v\sigma_{\theta\theta} - v\sigma_{zz}$$
$$E\varepsilon_{\theta\theta} = \sigma_{\theta\theta} - v\sigma_{zz} - v\sigma_{rr}$$
$$E\varepsilon_{zz} = \sigma_{zz} - v\sigma_{rr} - v\sigma_{\theta\theta}$$
$$G\gamma_{r\theta} = \tau_{r\theta}$$
$$G\gamma_{\theta z} = \tau_{\theta z}$$
$$G\gamma_{zr} = \tau_{zr}$$

Hence,

$$E\varepsilon_{rr} = \sigma_h(1-v) - v\sigma_v$$
$$E\varepsilon_{\theta\theta} = \sigma_h(1-v) - v\sigma_v$$
$$E\varepsilon_{zz} = \sigma_v - 2v\sigma_h$$
$$\gamma_{r\theta} = 0$$
$$\gamma_{\theta z} = 0$$
$$\gamma_{zr} = 0$$

where all shear strains are zero. Thus, $\varepsilon_{rr} = \varepsilon_{\theta\theta} = \varepsilon_h$ and

$$\sigma_h = \left(\frac{v}{1-v}\right)\sigma_v + E\varepsilon_h$$

Under the assumption of complete lateral restraint, the horizontal strain is zero and the result is the same as before (103).

5.3   Axial symmetry

When the problem at hand has an axis of rotational symmetry, reduction to a two-dimensional form is possible. The resulting system of equations is similar in some respects to the special cases of plane stress and plane strain. If shear stresses parallel to the axis of symmetry vanish, the problem becomes almost one-dimensional and amenable to analytical solution. Despite the simplifying assumptions of axial symmetry, the problem has many important practical applications. The analysis illustrates an approach to solving the entire system of equations formed from physical laws, kinematics, and Hooke's law.

In axial symmetry, all quantities including applied loads are first referred to a cylindrical system of coordinates r, θ, z and are then assumed to be independent of θ, so the z-axis is an axis of symmetry and the analysis may proceed in the rz-plane. The non-zero stresses are σ_{rr}, $\sigma_{\theta\theta}$, σ_{zz}, and τ_{rz}, while the zero shear stresses $\tau_{\theta r}$ and $\tau_{\theta z}$ imply zero shear strains $\left(\gamma_{\theta r}$ and $\gamma_{\theta z}\right)$ through Hooke's law. Displacements in the r- and z-directions are u and w, respectively. Displacement v in the θ-direction is zero because of problem symmetry. The strain–displacement equations are

$$\varepsilon_{rr} = \frac{\partial u}{\partial r}, \quad \varepsilon_{oo} = \frac{u}{r}, \quad \varepsilon_{zz} = \frac{\partial w}{\partial z}, \quad \gamma_{rz} = \frac{\partial u}{\partial z} + \frac{\partial w}{\partial r} \tag{105}$$

With the exception of the circumferential normal strain $\varepsilon_{\theta\theta}$, these equations are similar to the two-dimensional Cartesian form. Equilibrium in axial symmetry is expressed by

$$\frac{\partial \sigma_{rr}}{\partial r} + \frac{\partial \tau_{rz}}{\partial z} + \left(\frac{\sigma_{rr} - \sigma_{\theta\theta}}{r} \right) + \gamma_r = 0$$
$$\frac{\partial \tau_{rz}}{\partial r} + \frac{\partial \sigma_{zz}}{\partial z} + \frac{\tau_{rz}}{r} + \gamma_z = 0 \tag{106}$$

Stress equilibrium thus involves two equations and four unknowns, unlike the plane problem that involves two equations and three unknowns. As the radius r becomes indefinitely large, the curvature approaches zero and the plane case is recovered, as one would expect.

The most common axially symmetric problem in rock mechanics involves cylindrical laboratory test specimens obtained by diamond core drilling. In the field, the most common problem treated as the one of axial symmetry involves circular shafts, tunnels, and liners (Figure 5.2). An assumption often made in such problems is that in the region of interest, the shear stress τ_{rz} is zero. With this additional assumption, the equations of equilibrium reduce to

$$\frac{\partial \sigma_{rr}}{\partial r} + \left(\frac{\sigma_{rr} - \sigma_{\theta\theta}}{r} \right) + \gamma_r = 0$$
$$\frac{\partial \sigma_{zz}}{\partial z} + \gamma_z = 0$$

where the z-axis is chosen to coincide with the vertical gravity axis. The second equation can be integrated as before. Thus,

$$\sigma_{zz} = -\gamma z + f(r,\theta) = \gamma h + \sigma_o \tag{107}$$

where at $z = 0$, $\sigma_{zz} = \sigma_o$ is the externally applied load, and h is the depth with *compression* considered *positive*. In the laboratory problem, the gravity term is neglected, while in most field problems, the applied stress at the surface $\sigma_o = 0$. Interestingly enough, in the case of a circular shaft, as shown in Figure 5.2 (compression positive), the vertical stress before and after excavation is just the gravity stress, given by (107) with $\sigma_o = 0$. The vertical stress change $\Delta\sigma_{zz}$ induced by shaft excavation ("sinking") is zero.

The stresses before and after excavation must satisfy the stress equations of equilibrium and so must the stress changes. Equilibrium therefore requires

$$\frac{\partial \Delta\sigma_{rr}}{\partial r} + \left(\frac{\Delta\sigma_{rr} - \Delta\sigma_{\theta\theta}}{r} \right) = 0$$
$$\frac{\partial \Delta\sigma_{zz}}{\partial z} = 0$$

where Δ means change. The second equation is already satisfied, but the first equation is more readily solved in terms of displacement that occurs as a consequence of excavation. The reason is that only radial displacement occurs.

From Hooke's law in incremental form and the strain–displacement equations, one obtains

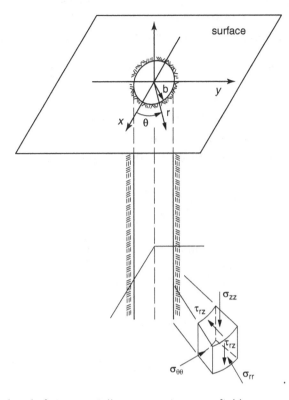

Figure 5.2 A circular shaft in an axially symmetric stress field.

$$\Delta\sigma_{rr} = \left(\frac{E}{1-v^2}\right)(\varepsilon_{rr} + v\varepsilon_{\theta\theta}) = \left(\frac{E}{1-v^2}\right)\left(\frac{\partial u}{\partial r} + \frac{u}{r}\right)$$

$$\Delta\sigma_{rr} - \Delta\sigma_{\theta\theta} = \left(\frac{E}{1+v}\right)(\varepsilon_{rr} - \varepsilon_{\theta\theta}) = \left(\frac{E}{1+v}\right)\left(\frac{\partial u}{\partial r} - \frac{u}{r}\right)$$

where the strains and displacements are associated with the induced stress changes. Substitution into the first equilibrium equation gives

$$\frac{\partial^2 u}{\partial r^2} + \left(\frac{1}{r}\right)\left(\frac{\partial u}{\partial r}\right) - \left(\frac{u}{r^2}\right) = 0$$

that has the general solution

$$u = Ar + B/r \tag{108}$$

where A and B are constants of integration to be determined by boundary conditions specific to the problem at hand. In the laboratory problem, the radial displacement is zero on the axis of symmetry, so B must be zero for this problem. In the field, the radial displacement induced by excavation must diminish to zero with distance from the excavated shaft wall shown in Figure 5.2, so A must be zero. However, in the case of a hollow cylinder (shaft liner), neither constant is zero. Interestingly, in case of a spherical cavity under hydrostatic stress, the displacement is also purely radial and is given by $u = Ar + B/r^2$. The first term accounts for initial stress (strain), while the second term accounts for excavation-induced changes.

In the shaft problem with $A = 0$, the strains caused by excavation are

$$\varepsilon_{rr} = -\frac{\partial u}{\partial r} = \left(\frac{B}{r^2}\right), \quad \varepsilon_{\theta\theta} = -\left(\frac{u}{r}\right) = -\left(\frac{B}{r^2}\right)$$

where the minus sign is introduced into the strain–displacement equations because compression is considered positive in this problem. The radial stress change from Hooke's law is

$$\Delta\sigma_{rr} = \left(\frac{E}{1-v^2}\right)\left(\frac{B}{r^2} - v\frac{B}{r^2}\right) = \left(\frac{EB}{1-v^2}\right)\left(\frac{1}{r^2}\right)$$

At the shaft wall where $r = b$, the radial stress change must be equal in magnitude, but opposite in sign to the preshaft radial stress because the post-excavation shaft wall is stress-free. Thus,

$$\sigma_{rr}|_{r=b} = (\sigma_{rr}^o + \Delta\sigma_{rr})|_{r=b}$$

Hence,

$$\Delta\sigma_{rr}|_{r=b} = -\sigma_{rr}^o|_{r=b} = -\sigma_h^o$$

and therefore,

$$B = \frac{-\sigma_h^o b^2(1-v^2)}{E}$$

$$\Delta\sigma_{rr} = -\sigma_h^o \frac{b^2}{r^2}$$

$$\Delta\sigma_{\theta\theta} = +\sigma_h^o \frac{b^2}{r^2}$$

Thus, after excavation

$$\sigma_{rr} = \sigma_h^o\left(1 - \frac{b^2}{r^2}\right)$$

$$\sigma_{v\theta} = \sigma_h^o\left(1 + \frac{b^2}{r^2}\right)$$

$$\sigma_{zz} = \sigma_v^o = \gamma h \tag{109}$$

$$\sigma_h^o = K\sigma_v^o$$

$$u = -\left(\frac{1-v^2}{E}\right)\sigma_h^o\left(\frac{b}{r}\right)^2$$

where the superscript o is added to emphasize preexcavation values. Inspection of these results shows that the radial stress is indeed zero at the shaft wall and that the stresses rapidly approach preexcavation values with distance from the shaft walls. These results also show that the circumferential stress is greatest at the shaft wall. Because K_o may be greater than one, either the circumferential or the vertical stress may be the major principal (compression) stress at the shaft wall. The negative sign on displacement indicates an inward radial motion is induced by excavation, as one would expect in an axially symmetric initial stress field.

Finally, because

$$\varepsilon_{zz} = v\left(\Delta\sigma_{rr} + \Delta\sigma_{\theta\theta}\right) = v\left(-\sigma_h^o \frac{b^2}{r^2} + \sigma_h^o \frac{b^2}{r^2}\right) = 0$$

for all r and z, no vertical strain is induced by excavation of the shaft. Consequently, there is no vertical displacement induced by excavation ($w = 0$). Of course, this is not the case near the shaft bottom, where the analysis does not apply.

6   LIMITS TO ELASTICITY – STRENGTH

Limits to purely elastic behavior are set by material strengths. In this context, strength is stress at the elastic limit. Strengths determined by uniaxial tension, compression, or possibly torsion are simply numbers, T_o, C_o, or R_o, as the case may be. These numbers mark the limit to a purely elastic response to load. Strength under biaxial or triaxial stress states may also be defined as the limit to a purely elastic response. However, a single number is no longer adequate to describe such strength. In three dimensions, strength is a function, not a single number. Instead of $\sigma - T_o = 0$, $\sigma - C_o = 0$, or $\sigma - R_o = 0$, one has $F(\sigma) = 0$, where σ stands for components of stress at the elastic limit, that is, $F(\sigma_{xx}, \ldots, \tau_{xy,\ldots}) = 0$. Constants that appear in the function F are material properties that characterize strength or simply "strengths".

Brittle materials, by definition, fail by fracture at the elastic limit with negligible inelastic deformation. Generally, tensile fracture, fracture under compressional load (shear fracture), and fracture under torsional stress (sometimes referred to as diagonal tension) lead to different strengths. Clearly, when tensile and compressive strengths are not equal, a single number is inadequate for the description of material strength. Thus, even under uniaxial stress, brittle materials require a functional description.

Ductile materials fail by yielding and experience plastic deformation as loading is continued beyond the elastic limit. This deformation is seen as a permanent "set" upon release of load. Figure 6.1 shows ideally brittle and ductile stress–strain plots under uniaxial tension. When the elastic limit of a material increases during yielding, the material *work hardens* or *strain hardens*, as shown in Figure 6.2. A falling stress–strain curve indicates *strain softening*.

Failure by ductile fracture often occurs after noticeable yielding. The original limit to elasticity of a ductile, strain hardening material is the *initial yield stress*, as shown in Figure 6.2. The high point of a uniaxial stress–strain curve is the *ultimate strength*, while the end point is the *rupture strength*. Unloading from a point in the elastic–plastic range of strain is elastic, as shown in Figure 6.2. Reloading is also elastic until the previous stress level is reached. Further plastic yielding is then possible.

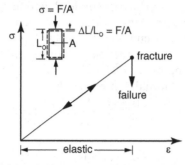

(a) ideally elastic – brittle failure

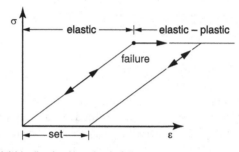

(a) ideally elastic – plastic failure

Figure 6.1 Ideally brittle (a) and ductile (b) stress–strain plots.

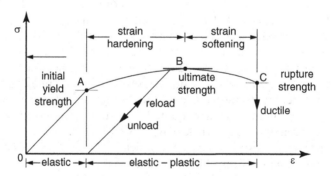

Figure 6.2 Ductile strain hardening, softening, and rupture.

Ductile materials often have equal tensile and compressive strengths, so strength in pure shear is just one half strength under normal stress failure. In three dimensions, the mode of failure may change. Materials that are brittle under uniaxial load may become ductile under triaxial loading and thus experience a brittle–ductile transition. Materials that are ductile at room temperature may become brittle under extreme cold. One concludes that a functional description of strength is needed in any case.

In consideration of Hooke's law and linearly elastic behavior, the uniaxial tensile stress failure criteria, $\sigma - T_o = 0$, $\sigma - C_o = 0$, or $\sigma - R_o = 0$ are also uniaxial strain failure criteria. For example, $\sigma - T_o = E\varepsilon - E\varepsilon_t = 0$, where E is Young's modulus and ε_t is the uniaxial tensile strain to failure. Thus, $\varepsilon - \varepsilon_t = 0$. Tensile strain to failure is often of the order of 0.1% or less, while strain to failure in uniaxial compression is of the order 1.0% for many rock types. More generally, any state of stress at the limit to elasticity is associated with a companion state of strain through Hooke's law. For this reason, there is no substantial difference between stress and strain failure criterion. One implies the other when strength is defined as stress at the elastic limit.

The most popular functions that describe strength in rock mechanics have names:

1. Mohr–Coulomb (MC),
2. Hoek–Brown (HB), and
3. Drucker–Prager (DP).

The functions these names imply are failure criteria, also referred to as yield functions. They apply to isotropic materials that lack directional features, but may be extended to anisotropic materials. Materials such as laminated sandstones that have different strengths parallel and perpendicular to the laminations are anisotropic and require more complex failure criteria than MC, HB, or DP. Because the strength of isotropic materials is independent of direction, the orientation of the principal stresses is not important. Hence, instead of $(\sigma_{xx}, \sigma_{yy}, \sigma_{zz}, \tau_{xy}, \tau_{yz}, \tau_{zx})$ in F or $(\sigma_1, \sigma_2, \sigma_3, \theta_x, \theta_y, \theta_z)$, where $(\sigma_1, \sigma_2, \sigma_3)$ are the major, intermediate, and minor principal stresses and the angles specify the direction of the 1-axis relative to xyz (the 1–2–3-axes from an orthogonal right-handed triple), one only needs magnitudes $(\sigma_1, \sigma_2, \sigma_3)$ to describe the strength of isotropic materials. Most strength functions are simple polynomials in stress, even in case of anisotropy. However, not all functions are candidates for the description of material strength; there are restrictions. One obvious restriction is that any such function gives the appropriate strength under uniaxial tensile, compressive, or shear loading. Another restriction is a single numerical value of strength in pure shear.

General three-dimensional stress states may be visualized in terms of the principal $(\sigma_1, \sigma_2, \sigma_3)$ acting on the faces of a small cube (principal planes). The principal directions define a mutually orthogonal triple of axes along which the principal stresses may be plotted. In such a plot, a is measured in the positive or negative 1-direction depending on algebraic sign and sign convention. The same is true of σ_2 and σ_3. As the stresses are varied, a set of points that define the limit to elasticity is generated. These points define a surface $F(\sigma_1, \sigma_2, \sigma_3) = 0$ that represents the relationship among the stresses at failure in principal stress space. The function F is an implicit failure criterion. With a change in variables emphasizing the plane of major and minor principal stresses

$$\sigma_m = (1/2)(\sigma_1 + \sigma_3), \quad \tau_m = (1/2)(\sigma_1 - \sigma_3), \quad \sigma = \sigma_2$$
$$\sigma_1 = (\sigma_m + \tau_m), \quad \sigma_3 = (\sigma_m - \tau_m), \quad \sigma_2 = \sigma$$

one has an explicit form $\tau_m = f(\sigma_m, \sigma)$. Other forms are possible, of course. For example, the inverse is $\sigma_m = g(\tau_m, \sigma)$. If the failure criterion is considered independent of the intermediate principal stress, then $\tau_m = f(\sigma_m), \sigma_m = g(\tau_m)$.

6.1 Mohr–Coulomb failure

The Mohr–Coulomb failure criterion is

$$|\tau| = \sigma \tan(\phi) \tag{110}$$

where τ and σ are shear and normal stresses at a point on a potential failure surface and compression is positive. The angle ϕ is the *angle of internal friction*; c is the *cohesion*. These properties characterize the strength of the material. In this regard, the algebraic sign of the shear stress is not physically significant, so the absolute value sign is used. Mohr strength theory postulates a functional relationship between τ and σ, that is $\tau = f(\sigma)$. A linear form is associated with Coulomb [ref]. This criterion is a pair of straight lines when plotted in a normal stress (x-axis)–shear stress (y-axis) plane, as shown in Figure 6.3. The slopes of these lines is $\pm\tan(\phi)$; they have r-axis intercepts $\pm c$. The function (MC) is thus symmetric with respect to the a-axis and is tangent to Mohr's circles representing stress states at failure. Function (MC) is sometimes referred to as the Mohr envelope.

The MC criterion may also be expressed in terms of the major and minor principal stresses. Thus,

$$\left| \left(\frac{\sigma_1 - \sigma_3}{2} \right) \right| = \left(\frac{\sigma_1 + \sigma_3}{2} \right) \sin(\phi) + (c)\cos(\phi) \tag{111}$$

Alternatively,

$$|\tau_m| = (\sigma_m)\sin(\phi) + (c)\cos(\phi) \tag{112}$$

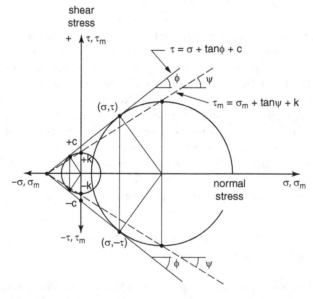

Figure 6.3 Mohr–Coulomb failure envelope and Mohr's stress circles at failure; the circle centered at the origin represents failure in pure shear.

where τ_m is the maximum shear stress (also the numerical value of the minimum shear stress) and σ_m is the mean normal stress in the plane of σ_1, σ_3. This function is also a pair of straight lines in the normal stress–shear stress plane, as seen in Figure 6.3. These lines pass through the tops $(\sigma_m, +\tau_m)$ and bottoms $(\sigma_m, -\tau_m)$ of the Mohr circles representing stress states at failure.

A simple generalization is

$$|\tau_m| = (\sigma_m)\tan(\psi) + k \tag{113}$$

where the slope $\tan(\psi)$ may now exceed the limits of ± 1 set by $\sin(\phi)$. An envelope exists whenever ψ is less than $\pm\pi/4$. In the unusual case that a failure envelope does not exist, this last form may be used. Shear axis intercepts of this form of MC are $|\tau_m| = (c)\cos(\phi)$, which gives strength in pure shear with the name Tresca. The Tresca or maximum shear stress criterion is often used for ductile metals plane, as shown in Figure 6.4. Shear axis intercepts are $\tau_m = \pm k$, the Tresca strength in pure shear. A Mohr's circle representing the stress state at failure has a constant radius $|\tau_m| = k$ that is independent of the mean normal stress. The Tresca criterion may also be considered a maximum shear strain to failure criterion in consideration of Hooke's law and linearly elastic behavior up to the elastic limit.

6.2 Hoek–Brown failure

The original Hoek–Brown failure criterion may be stated as

$$\sigma_1 = \sigma_3 + \sqrt{a\sigma_3 + b^2}$$

where a and b are strength properties of the material. This criterion is obviously non-linear. In, τ_m, σ_m terms (HB) is

$$\tau_m = \left(\frac{1}{2}\right)\left[a(\sigma_m - \tau_m) + b^2\right]^{1/2} \tag{115}$$

or

$$(\tau_m)^2 = \left(\frac{a}{4}\right)\left[(\sigma_m - \tau_m) + b^2\right] \tag{116}$$

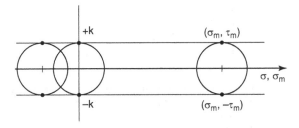

Figure 6.4 Tresca failure criterion and Mohr's stress circles at failure.

that after solving for τ_m gives

$$\tau_m = -\left(\frac{a}{8}\right) \pm \sqrt{\left(\frac{a}{8}\right)^2 + \left(\frac{1}{4}\right)\left(a\sigma_m + b^2\right)} \tag{117}$$

which is puzzling because when $\sigma_m = 0$, two substantially different values of strength in pure shear are implied. The values are the shear axis intercepts given by

$$\tau_m = -\left(\frac{a}{8}\right) \pm \sqrt{\left(\frac{a}{8}\right)^2 + \left(\frac{b^2}{4}\right)} \tag{118}$$

The constant a is usually an order of magnitude larger than b, so the intercepts are very different. Neither of these latter forms of HB are symmetric about the normal stress axis in the normal stress–shear stress plane, as seen in Figure 6.5. Thus, HB requires further consideration before use in practical applications. A simple fix would be to restrict HB to

$$|\tau_m| = -\left(\frac{a}{8}\right) + \sqrt{\left(\frac{a}{8}\right)^2 + \left(\frac{1}{4}\right)\left(a\sigma_m + b^2\right)}, \quad \sigma_m \geq -\frac{b^2}{a} \tag{119}$$

The absolute value sign insures symmetry with respect to the normal stress axis, while the restriction on the mean normal stress simply requires the mean normal stress to be greater than the normal stress axis intercept. This value of the restriction is approximately equal to tensile strength $(-T_o)$ for many rock types that have ratios of compressive to tensile strength of 10 or greater. Restricted HB in (119) is a pair of nearly straight lines that come to a point on the normal stress axis, much like MC.

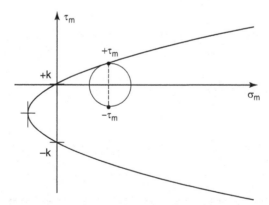

Figure 6.5 Hoek–Brown failure criterion showing two shear strengths and a Mohr's stress circle at failure with respect to positive shear stress. The criterion is not symmetric with respect to the normal stress axis and thus shows two shear strengths of different values depending on whether the shear stress is positive or negative.

6.3 Drucker–Prager failure

The Drucker–Prager failure criterion expressed in terms of the principal stresses is

$$\left(\sqrt{\frac{2}{3}}\right)\left[\left(\frac{\sigma_1-\sigma_2}{2}\right)^2+\left(\frac{\sigma_2-\sigma_3}{2}\right)^2+\left(\frac{\sigma_3-\sigma_1}{2}\right)^2\right]^{1/2}=A(\sigma_1+\sigma_2+\sigma_3)+B \qquad (120)$$

where A and B are strength properties of the material and compression is considered positive. The term on the left of DP equation is J_2 times the root-mean-square value of the principal shears because the terms under the square root sign are the squares of the principal shear stresses (relative maxima and minima). The first term in parentheses on the right is just three times the three-dimensional mean normal stress. This criterion is often written in abbreviated form:

$$J_2^{1/2} = AI_1 + B \qquad (121)$$

where J_2 is the second principal invariant of deviatoric stress and I_1 is the first principal invariant of total stress. In pure shear, DP strength is $|\tau_m| = B$.

The DP criterion is a cone centered on the space diagonal when plotted with the principal stresses $\sigma_1, \sigma_2, \sigma_3$ as coordinates, as shown in Figure 6.6. When the strength parameter $A = 0$, DP reduces to a criterion associated with von Mises that is widely used for ductile metals (τ_{oct}) because $\sqrt{J_2}$ is a constant times the octahedral shear stress $\left(\tau_{oct} = \sqrt{(2/3)J_2}\right)$; the octahedral plane is normal to the space diagonal in principal stress space. The von Mises $\left(\sqrt{J_2} = B\right)$ and octahedral shear stress criteria are cylinders centered on the space diagonal in principal stress space. The space diagonal is also referred to as the *hydrostatic axis* because points on the diagonal represent hydrostatic states of stress characterized by equality of the three principal stresses.

The DP strength criterion differs from MC and HB by inclusion of the intermediate principal stress, although it is linear in stress. The MC and HB criteria imply that the intermediate principal stress has only a negligible influence on strength. The MC criterion forms a pyramidal surface in principal stress space with flat sides, as shown

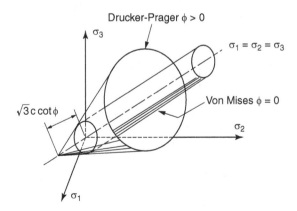

Figure 6.6 Von Mises and Drucker–Prager failure criteria in stress space. Compression is positive.

in Figure 6.7. The HB criterion would also form a pyramid-like surface with rounded sides, but would not be centered on the space diagonal.

Nonlinear forms of MC and DP are possible. Thus,

$$|\tau|^n = \sigma \tan(\phi) + c \tag{122a}$$

$$|\tau_m|^n = \sigma_m d + e \tag{122b}$$

$$(J^{1/2})^N = AI_1 + B \tag{122c}$$

where n and N are some positive numbers, not necessarily integers, and d and e are the strength parameters. These *n- and N-type failure criterion* reduce to MC, extended MC, and DP, respectively, when $n = 1$ or $N = 1$. HB is nonlinear at the outset. By changing the square root sign to some other value than the corresponding ½ exponent, a different degree of nonlinearity may be achieved that better fits experimental data. Indeed, the value of the exponent n or N in any of failure criteria must be determined by experiment. Although there is no theoretical limit to n or N, a value between one and two is likely in consideration of laboratory test data. All of the strength parameters introduced with MC, DP, and HB can be expressed in terms of the uniaxial tensile and unconfined compressive strengths, T_o and C_o. Because the considered materials are isotropic, shear strength R_o may also be expressed in terms of T_o and C_o. Thus,

$$
\begin{aligned}
&\sin(\phi) = \frac{C_o - T_o}{C_o + T_o}, && C_o = \frac{2c\sin(\phi)}{1 - \sin(\phi)}, && R_o = \frac{C_o T_o}{C_o + T_o} \\
&c = \sqrt{C_o T_o / 4} && T_o = \frac{2c\sin(\phi)}{1 + \sin(\phi)}, && \\
&a = \frac{C_o^2 - T_o^2}{T_o} && b = C_o, && T_o = -\frac{a}{2} + \frac{1}{2}\sqrt{a^2 + 4b^2} \\
&A = \frac{1}{\sqrt{3}}\left(\frac{C_o - T_o}{C_o + T_o}\right) && C_o = \frac{\sqrt{3}B}{1 - \sqrt{3}A} && R_o = \frac{2}{\sqrt{3}}\left(\frac{C_o T_o}{C_o + T_o}\right) \\
&B = \frac{2}{\sqrt{3}}\left(\frac{C_o T_o}{C_o + T_o}\right) && T_o = \frac{\sqrt{3}B}{1 + \sqrt{3}A} &&
\end{aligned}
\tag{123}
$$

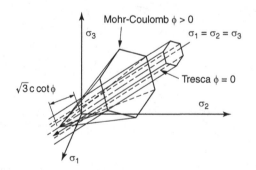

Figure 6.7 Tresca and Mohr–Coulomb failure criteria in principal stress space. Compression is positive.

The pure shear strengths of DP are computed when $\sigma_3 = -\sigma_1$, $\sigma_2 = 0$. Corresponding expressions may be obtained for n-type criterion in the same way, by solving for the two unknown strength constants in terms of unconfined compressive and tensile strengths.

6.4   Compressive strength under confining pressure

General three-dimensional stress states are difficult to obtain experimentally and require specialized laboratory test apparatus. The usual test procedure uses cylindrical test specimens that are axially loaded and laterally confined. The axial load $\sigma_{zz} = \sigma_1$, while the lateral confining pressure is σ_3. Under these conditions, the radial strain is equal to the circumferential strain, so in view of the assumed linear elasticity, the radial and circumferential stresses are also equal. Thus, $\sigma_2 = \sigma_3 = \sigma_{rr} = \sigma_{\theta\theta}$. These conditions define the popular triaxial strength test that determines compressive strength under confining pressure.

Each of the three strength criteria, MC, HB, and DP, can be specialized for the triaxial strength test. Each shows an increase in compressive strength with increasing confining pressure. However, the rate of increase differs from criterion to criterion. Rearrangement of MC shows that

$$\sigma_1 = \sigma_3 \left(\frac{1 + \sin(\phi)}{1 - \sin(\phi)} \right) + \left(\frac{2c \cos(\phi)}{1 - \sin(\phi)} \right) \tag{124}$$

Alternatively,

$$\sigma_1 = C_o + \left(\frac{C_o}{T_o} \right) p \tag{125}$$

Hence, compressive strength C_p under confining pressure p according to MC is

$$C_p = C_o + \left(\frac{C_o}{T_o} \right) p \tag{126}$$

with rate of increase of strength given by

$$\left(\frac{dC_p}{dp} \right) = \left(\frac{C_o}{T_o} \right) \tag{127}$$

A similar analysis shows that compressive strength under confining pressure according to HB is

$$C_p = p + \sqrt{\left(\frac{C_o^2 - T_o^2}{T_o} \right) p + C_o^2} \tag{128}$$

with rate of increase approximately

$$\frac{dC_p}{dp} = \frac{C_o}{2T_o} + 1 \tag{129}$$

Compressive strength under confining pressure according to DP is

$$C_p = C_o + \left(\frac{3C_o}{2T_o} - \frac{1}{2} \right) p$$

(130)

with rate of increase

$$\frac{dC_p}{dp} = \frac{3C_o}{2T_o} - \frac{1}{2}$$

(131)

The ratio of unconfined compressive to tensile strength for rock is typically in the range of 10–20. In this range, HB shows an increase in compressive strength with confining pressure about 50% less than MC, while DP shows an increase of about 50% more. These forecasts seem to have never been tested experimentally.

Strength test data plotted in σ_3 (x-axis)–σ_1 (y-axis) plane often show a steeply rising curve on the tensile side of the axis that becomes less steep on the compressive side. Fitting as straight line to triaxial strength data over a limited range may produce a reasonable estimate of unconfined compressive strength C_o when the fitted line is extrapolated to the σ_1-axis. However, because the slope of the fitted line, dC_p/dp, is often significantly less than that given by the preceding expressions involving C_o and T_o, continued extrapolation to the σ_3 is likely to produce an inaccurate estimate of T_o, an overestimate. A question then arises whether to use unconfined compressive and tensile strength test data, C_o and T_o, or slope and intercept data, for example, c and $\tan(\phi)$ in MC, in analyses. One response to this question would be to use a nonlinear expression from (122) to fit data. Another would be to retain the simplicity of a linear fit, but restrict analysis to a range of stress states bounded by experimental data.

Figure 6.8 shows laboratory test data including tensile strength, unconfined compressive strength, and compressive strength determined under confining pressure. MC,

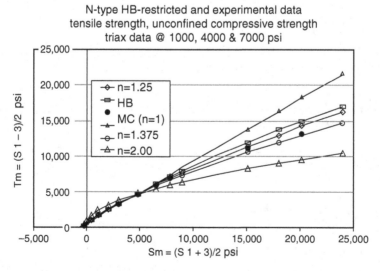

Figure 6.8 Comparisons of MC, restricted HB, and n-type failure criteria with experimental test results on a sandstone.

restricted HB, and n-type criteria are also plotted. Four values of n are used, which range between 1.00 and 2.00. The best fit to the experimental data is obtained using the n-type criterion (122b) with $n = 1.375$. All plots pass through the point corresponding to unconfined compressive strength, C_o. The n-type plots also pass through the point corresponding to T_o, while HB approximates this point.

The data points in Figure 6.8 are averages of two or more tests at each confining pressure and many more unconfined compressive and tensile strength repetitions. This plot procedure masks scatter in the data, but allows for easier visual comparisons. The data in Figure 6.8 also show that linear extrapolation from unconfined compressive and tensile strength data to well into the range of confining pressures results in an overestimate of compressive strength when the failure envelope is curved or tends to decrease in slope with confining pressure, as is usually the case.

Figure 6.9 shows the confining pressure data and associated Mohr's circles. Inspection of Figure 6.9 shows clearly that backward extrapolation of confining pressure test data would overestimate tensile strength by a considerable amount and to a lesser degree overestimate unconfined compressive strength of this rock (sandstone).

6.5 Energy and stability

Energy may also be considered as a criterion for failure. For example, strain energy density U at the elastic limit may be postulated as a failure criterion. Under uniaxial stress, strain energy per unit volume of material $U = \sigma\varepsilon/2$ in case of normal stress; $U = \tau\gamma/2$ in case of shear stress. Alternatively, $U = \sigma^2/2E = E\varepsilon^2/2$, where E is Young's modulus and σ is T_o or C_o (ε is T_o/E or C_o/E), or $U = \tau^2/2G = G\gamma^2/2$, where G is a shear modulus and $\tau = R_o$ (γ is τ/G). Because strains and strain energies can be calculated from the stresses at failure through Hooke's law, no significant differences

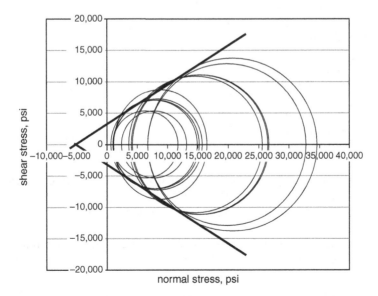

Figure 6.9 Mohr's circles and envelope for compressive strength under confining pressure.

exist between stress, strain, or strain energy criteria for failure when strength is defined as stress at the elastic limit. The same is true in three dimensions.

A more subtle limit to purely elastic behavior than strength is *stability*. Stability and instability may be considered globally as in structural mechanics and elastic buckling of columns or locally as in the onset of fast fracture of brittle materials. Both result in a sudden large displacement toward a new equilibrium position with little increase in load. The new equilibrium position may be unattainable, so the end result is collapse of the considered body. Stability analysis is usually done through an energy approach to a region initially at rest that lacks kinetic energy. A stationary value of energy is a necessary condition for instability. If energy is a minimum, then equilibrium is stable; a maximum of energy is unstable.

Griffith supposed that stress concentration at the tips of minute flaws or microcracks within a solid led to macroscopic fracture according to an energy criterion. Behavior of these *Griffith cracks* explained why Young's modulus and strengths of crystalline solids were orders of magnitude less than the results from theoretical calculations of chemical bond strength in a lattice. In the case of failure by fracture, surfaces along the fracture absorb energy from the surrounding material and applied loads, although the opposing surfaces are not in frictional contact. An elliptical opening of high aspect ratio formed a mathematical model of a Griffith crack in two dimensions. A penny-shaped crack is often used in three-dimensional analyses.

The energy of a quasi-static system with fractures or cracks is

$$E = (-W_L + U_E) + U_S$$

where W_L, U_E, and U_S are the potential energy of the applied loads, strain energy of the deformed system, and surface energy of fracture. Slow application of the loads to a linearly elastic material leads to $W_L = 2U_E$; hence, the energy balance under constant applied loads is $E = -U_E + U_S$.

The time rate of energy change is

$$\frac{dE}{dt} = \frac{dW_L}{dt} + \frac{dU_E}{dt} + \frac{dU_S}{dt}$$

$$= \left(-\frac{dW_L}{dA} + \frac{dU_E}{dA} \right)\left(\frac{dA}{dt} \right) + \frac{dU_S}{dt}$$

$$= -G\left(\frac{dA}{dt} \right) + \gamma\left(\frac{dA}{dt} \right)$$

$$\frac{dE}{dt} = (-G + \gamma)\left(\frac{dA}{dt} \right)$$

where A is the fracture surface area (ft^2, m^2), γ is the fracture surface energy per unit area (ft-lbf/ft^2, N-m/m^2), and G is the *fracture extension force* that is defined as the derivative of the mechanical energy release per unit area of fracture (ft-lbf/ft^2, N-m/m^2). The right-hand side of the last equation is, thermodynamically, an intensive variable times a change in an extensive variable (per unit of time) and is thus a differential of work. Thermodynamic equilibrium requires the time rate of energy change to be zero. Hence,

$$\left(-G + \gamma \right)\left(\frac{dA}{dt} \right) = 0$$

is the requirement for continuing equilibrium; the crack extension force must be just balanced by surface energy of the fracture. When equilibrium corresponds to an energy minimum, then excursions from equilibrium are associated with positive energy changes, that is, $dE>0$, as seen in Figure 6.10a. If $G>\gamma$, then dA must be negative, implying a fracture area decrease or *healing*. At an energy maximum, energy excursion from equilibrium is negative, $dE<0$, as seen in Figure 6.10b. If $G<\gamma$, then dA must be positive, implying a fracture *growth*. Thus, fractures may grow under constant applied loads and displacements.

The fracture extension force G is related to the intensity of stress at the fracture ends. There are three basic types of *stress intensity factors* identified in fracture

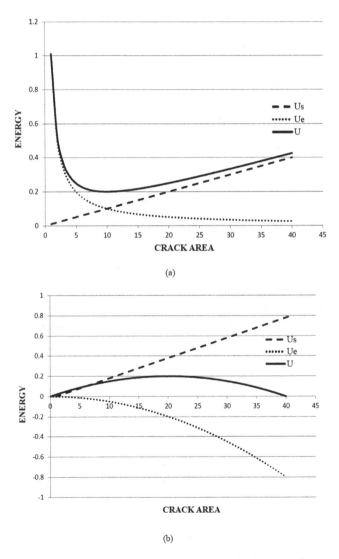

Figure 6.10 Energy minimum (a) and maximum (b) for a cracked solid.

mechanics. These factors are K_I, K_{II}, and K_{III} and are associated with tensile and shear fractures illustrated in Figure 6.11. Stress intensity factors depend on the applied loads and fracture size and may be used to characterize material toughness or resistance to fractures. Specialized laboratory test procedures are used to determine fracture toughness or stress intensity factors, of which especially important is K_{Ic}.

Addition of a kinetic energy term to the energy balance gives

$$E = (-W_L + U_E) + U_S + U_K$$

Thermodynamic equilibrium then requires

$$\frac{dE}{dt} = -(G - \gamma)\left(\frac{dA}{dt}\right) + \frac{dU_K}{dt}$$

Hence, dynamic fracture must be consistent with

$$\frac{dU_K}{dt} = (G - \gamma)\left(\frac{dA}{dt}\right)$$

Fast fracture growth may be accompanied by an increase in kinetic energy of the system.

An important result from fracture mechanics is that the rate of fracture growth is proportional to fracture size. Large cracks and fractures tend to grow faster. In an array of cracks, a crack that is initially somewhat longer than the other cracks of similar orientation quickly becomes dominant, while the smaller cracks tend to remain small. The larger radial cracks extending in star-like fashion from a cylindrical cavity grow preferentially to the shorter cracks.

Another important result is that below a critical size, cracks tend not to grow, but once a critical size or length is reached, relatively rapid growth follows. Crack growth may still occur at sub-critical size because of environmental effects at the crack tip that reduce the surface energy of the material. This effect may be especially noticeable in humid and corrosive environments. After a prolonged period of sub-critical crack growth, crack size may become critical followed by rapid crack growth and instability.

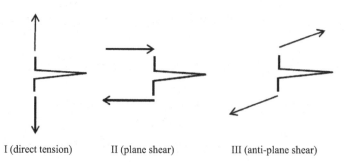

I (direct tension)　　　II (plane shear)　　　III (anti-plane shear)

Figure 6.11 Three types of fracture modes.

6.6   A statistical strength model

The numerous imperfections at grain boundaries and within grains and the great contrast in properties of minerals within a given rock type, for example in a granite, are candidate Griffith cracks. Growth of micro-cracks from such defects ultimately leads to macroscopic failure. Numerous laboratory studies of nominally intact test specimens confirm this phenomenon. A random distribution of flaw size and orientation in a test specimen leads to a statistical theory of strength associated with Weibull.

Consider a direct-pull tensile test, and suppose the test specimen of length L and cross-sectional area A is divided into a *series of n pieces* of length l such that $nl = L$ or $nv = V$, where v and V are subdivision volume and total volume, respectively. If the probability of failure of any one of the subdivisions is $P_o(\sigma)$ under tension σ, then the probability of any subdivision not failing is $1 - P_o(\sigma)$ and the probability of the test specimen as whole not failing is

$$1 - P = (1 - P_o)^n$$

The probability of failure of an infinitely long specimen $(n \to \infty)$ is 1, so the right-hand side goes to zero as n becomes indefinitely large. This is to say that a very long chain is certain to have a link that does not survive under the applied stress.

If the basic probability is such that

$$P = 1 - \exp[-f(\sigma)]$$

then the probability of failure is

$$P = 1 - \exp[-nf(\sigma)] = 1 - \exp[-(V/v)f(\sigma)]$$

The mean stress at failure is then

$$\bar{\sigma} = \int_0^1 \sigma\, dP = -\int_0^\infty \sigma \frac{d(1 - P_o)^n}{d\sigma} = \int_0^\infty (1 - P)\, d\sigma$$

where the last integral comes from an integration by parts. Thus,

$$\bar{\sigma} = \int_0^\infty \exp[-(V/v)f(\sigma)]\, d\sigma$$

which indicates that the mean tensile strength is inversely proportional to the ratio of test specimen size to grain size.

The statistical model of tensile strength thus leads to a *size effect* that implies lower strength of larger test specimen because of the greater probability of the test specimen containing a larger flaw. Other factors being equal, a smaller grain size increases strength. The dependency on the stress distribution indicates that different tensile strengths may be obtained under different test conditions, for example in bending rather than in direct pull.

One may suppose the same effects are true for failure in compression, although the weakest link model underlying the direct-pull tensile test is not apt for compressive failure. In compression testing of intact rocks, subdivision may form parallel elements and even series–parallel combinations. Statistical strength analysis of such models then become quite complex as failed elements transfer stress to adjacent elements without causing failure of the test specimen as a whole. As the micro-crack network grows and cracks join, a macroscopic fracture may be formed that is unstable and leads to test specimen failure.

6.7   Strain beyond the elastic limit

The onset of inelastic deformation is usually indicated by a large increase in strain with little increase in stress. When strain occurs concurrently with stress beyond the elastic limit, the plastic response is independent of time just as is the elastic part of the response. A work hardening material is intrinsically stable because an increase in stress causes a proportional strain increment that increases material strength. An incremental stress–strain relationship is favored because of nonlinearity. Thus, in this elastic–plastic range in one dimension,

$$d\sigma = (E - E_p)d\varepsilon$$

where E_p is a plastic correction to the elastic modulus (Young's modulus). This formulation is generalized for application to three-dimensional analysis.

Strain softening may be unstable because of strength reduction with increasing strain. In laboratory testing of intact rocks, micro-crack growth causes a progressive, but local loss of cohesion. However, the accumulation of damage may not cause rupture of the test specimen as a whole, despite the reduction in strength. Often, an experimental force–displacement curve shows an elastic rise to a *peak strength* that is followed by a sharp decline to a *residual strength*, as shown in Figure 6.12. Whether strain softening should be treated as a consequence of instability or as a strength characteristic of the material is an unsettled question. Practical considerations favor a strength description.

The simplest time-dependent response beyond the elastic limit may be formulated as a viscous or rate-dependent response such that stress depends on strain rate. A total strain rate is then composed of an elastic part obtained through Hooke's law and a viscoplastic part. Thus, in one dimension,

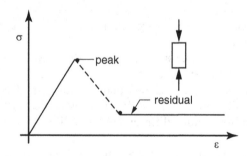

Figure 6.12 Peak residual strengths under compression.

$$\dot{\varepsilon} = \dot{\varepsilon}_e + \dot{\varepsilon}_p = \frac{\dot{\sigma}}{E} + \frac{\sigma - \sigma_o}{\eta}$$

where E is Young's modulus, η is the solid viscosity, and σ_o is the elastic limit or strength of the material. This response is that of a *Bingham* plastic solid. When stress is less than strength, the response is purely elastic, but only the excess stress above the elastic limit causes plastic strain. Generalization to three dimensions is possible.

Appendix B

Study questions

GEOFEM I

1. What is a finite "element"?
2. What is the "finite element method"?
3. Briefly explain the popularity of the finite element method.
4. What is meant by the "motion" of a material body?
5. What are the three main categories of equations that combine to allow one to compute the motion of a body?
6. Write Hooke's law in cylindrical coordinates (r, θ, z) for an isotropic material with Young's modulus E, Poisson's ratio v, and shear modulus G.
7. What is strain energy and how does it relate to the first law of thermodynamics (energy balance)?
8. Consider a computer program subroutine $NMTX$ (N-matrix) that computes interpolation functions $N_1(x, y)$, $N_2(x, y)$ for a triangular element shown in Figure 2.1 of the "Notes" that for any unknown such as temperature T that varies linearly with position (x, y) in the triangle; that is,

$$T = a_o + a_1 x + a_2 y$$

is also given in terms of the known temperatures T_1, T_2, T_3 at the corner points (nodes) by

$$T = N_1 T_1 + N_2 T_2 + N_3 T_3$$

Derive expressions for the interpolation functions, the Ns, and show that indeed they are linear functions of x and y.
9. Show that in the derivation of the interpolation functions in Problem 7, the determinant of the coefficients computed in solving for the as is twice the area of the considered triangle.
10. Show that the average of an interpolated node quantity such as temperature, pressure, or displacement component at the center (x_c, y_c) of a linear triangle is just the arithmetic average of the node values. Thus,

$$T(x_c, y_c) = (1/3)(T_1 + T_2 + T_3) = (1/A) \int_A T(x, y) \, dA$$

where A is the area of the triangle which is actually a triangular slab one unit thick and $dV = (1)dA$, V = volume. Note: The center coordinates are

$$x_c = (1/A) \int_A x \, dA \text{ and } y_c = (1/A) \int_A y \, dA$$

GEOFEM 2

1. Consider the tetrahedron as shown in the sketch.

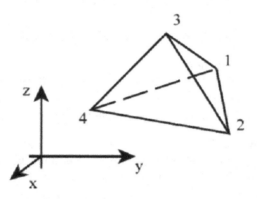

Assume a linear interpolation of displacements u, v, w in three dimensions x, y, z. Derive expressions for the interpolation functions (Ns) analogous to the two-dimensional case for a triangle. Use matrix notation for symbolic calculations, and then derive associated expressions for the six strains in three dimensions. Note: Element identification by nodes is 1, 2, 3, 4, obtained by looking down the dotted edge from node 4 to the opposing face and then listing the face nodes in counter-clockwise order. The triangular face, if it were a triangular element, would be identified by nodes 1, 2, 3 in counter-clockwise order. If the order is reversed, then during computations a sign reversal occurs that is undesirable.

2. Consider the triangle in the sketch below, then use the vector (cross) product to derive an expression for the area of the triangle. Show that the determinant of the coefficients is equal to twice the area of the triangle using notation from the course notes.

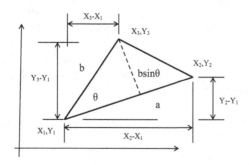

GEOFEM 3

1. Consider a quadrilateral element shown in the sketch.

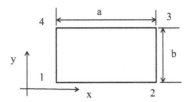

Assume a linear interpolation of temperature T in two dimensions, x and y, so that

$$T = A + Bx + Cy + Dxy$$

Shift the origin of coordinates to point 1, then derive the function for the rectangle, and bring into the form

$$T = N_1T_1 + N_2T_2 + N_3T_3 + N_4T_4$$

Note: $T(0,0) = T_1$, $T(a,0) = T_2$ $T(a,b) = T_3$, $T(0,b) = T_4$. Derive expressions for the interpolation functions (Ns), for example $N_3 = (xy)/(ab)$.

2. Suppose the quadrilateral in the sketch has node coordinates and temperatures

$T_1 = 100°$, x_1y_1: (2, 2) $T_{12} = 60°$, x_2y_2: (4, 2)
$T_3 = 50°$, x_3y_3: (4, 3) $T_4 = 90°$, x_4y_4: (2, 3)

Using the results from Problem 1, find the temperatures at (3, 1.5). What is the temperature at the center of the element and what are the coordinates there?

3. Sketch the temperature distribution above the xy-plane ($T=z$ coordinate over the element in Problem 2).

4. Determine the B-matrix for the quadrilateral shown in the sketch, so

$$\begin{Bmatrix} \varepsilon_{xx} \\ \varepsilon_{yy} \\ \gamma_{xy} \end{Bmatrix} = [B] \begin{Bmatrix} u_1 \\ \cdots \\ u_4 \\ v_1 \\ \cdots \\ v_4 \end{Bmatrix} = [B]\{\delta\}$$

GEOFEM 4

1. Consider a generic quadrilateral element shown in the sketch.

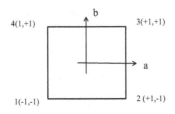

Consider the origin of coordinates at the center. Assume a linear interpolation, for example

$$u = \alpha_o + \alpha_1 a + \alpha_2 b + \alpha_3 ab \qquad (1)$$

where u is the b-component of displacement. Show algebraically that (1) can be brought into the form

$$u = N_1 u_1 + N_2 u_2 + N_3 u_3 + N_4 u_4 \qquad (2)$$

2. Derive expressions for the interpolation functions in (2) such that

$$N_1 = (1/4)(1-a)(1-b), \quad N_2 = (1/4)(1+a)(1-b),$$
$$N_3 = (1/4)(1+a)(1+b), \quad N_4 = (1/4)(1-a)(1+b)$$

Hint: Do the derivation directly from (1) by substituting numerical values of the nodes into (1) and then solving for the αs. An example is $\alpha_o = (1/4)(u_1 + u_2 + u_3 + u_4)$.

3. Consider the isoparametric quadrilateral shown in the sketch below, where $x = S_1 x_1 + S_2 x_2 + S_3 x_3 + S_4 x_4$ and similarly for y.

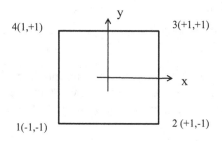

4(1,+1)   3(+1,+1)

y

x

1(-1,-1)   2 (+1,-1)

Show that

$$S_1 = (1/4)(1-x)(1-y), \quad S_2 = (1/4)(1+x)(1-y),$$
$$S_3 = (1/4)(1+x)(1+y), \quad S_4 = (1/4)(1-x)(1+y) \quad \text{and hence}$$

$x = (x_1/4)(1-x)(1-y) + (x_2/4)(1+x)(1-y) + (x_3/4)(1+x)(1+y) + (x_4/4)(1-x)(1+y)$ and similarly for y. Verify the center is at $(0, 0)$.

4. Consider a one-dimensional truss bar element shown in the sketch. Assume a linear interpolation of displacement u along the bar in the x-direction such that $u = a_o + a_1 x$. Derive the Ns so that $u = N_1 u_1 + N_2 u_2$, and then find a B-matrix that gives strain in terms of the node displacements (δ_1, δ_2) at the left- and right-hand sides of the bar. Finally, assuming the bar area is A and that tensile forces are applied at the bar ends, determine the bar stiffness matrix in the linearly elastic case, i.e., find $\{f\} = [k]\{\delta\}$ in this one-dimensional case.

   Hint: Start with Hooke's law for this last part.

x_1 ▬▬▬▬▬▬▬ x_2

⟶ x

GEOFEM 5

Code validation is an essential step toward development of a reliable numerical analysis technique. The simplest validation step is to compare solutions to known problems with results from a new or unfamiliar finite element computer program. As an example, a solution to the problem of a tip-loaded cantilever beam with finite element method (FEM) results is desired. The beam bending solution may be achieved using technical beam bending theory, that is, the flexure formula and the Euler–Bernoulli bending equations. Both may be found in mechanics of materials textbooks. This comparison is just the first step in a series of such comparisons.

In addition to code validation, this exercise also concerns behavior of constant-strain triangles (CSTs) and quadrilaterals composed of four constant-strain triangles (4CSTs) under bending stress. Also of interest is a comparison of Gauss elimination equation solving with Gauss–Seidel iteration.

A cantilever beam is shown in the figure that has a span of 60 in., a depth of 6 in., and a breadth of 1 in. into the page. The span depth ratio is 10, so beam action is assured. A point load is applied at the upper right-hand corner, while the left-hand side of the beam is fixed.

1. Consider an elastic analysis of this beam using the flexure and bending formulas that can be found in any mechanics of materials textbook. In particular, derive formulas for the bending stress (x-direction normal stress) as a function of position in the beam and for the deflection (y-displacement). Use a Young's modulus $E = 9.6 \times 106$ psi and a Poisson's ratio $v = 0.20$. Because the FEM approach is plane strain, analysis of this beam should also be plane strain. Thus, in place of E in the usual (plane stress) beam formulas, one should use $E/(1 - v^2)$.

   Three finite element meshes of the beam in Figure 1 are shown in Figure 2. There are 360 elements ($1 \times 1 \times 1$ in.) and 427 nodes in the quadrilateral mesh. There are 60 horizontal subdivisions and 6 vertical subdivisions in Figure 2a. Thus, the elements are unit squares in Figure 2a. Origin of coordinates is at the lower left-hand corner. Node number 1 is at the origin; numbering proceeds from left to right starting at the bottom and proceeding to the top. Elements are numbered in the same way.

   The point load is applied at node 427, the upper right-hand corner in Figure 2a. Nodes 1, 62, 123, 184, 245, 306, and 367 are at the left-hand side of the beam and are fixed in the x-direction. Additionally, the center node at the left-hand side,

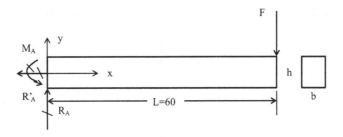

Figure 1 GEOFEM 5 (tip-loaded cantilever beam).

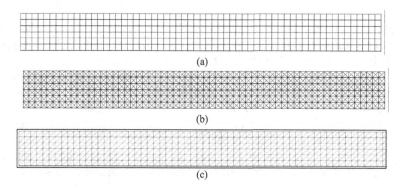

Figure 2 GEOFEM 5: (a) quadrilaterals – 4CSTs; (b) triangles (4); (c) triangles (2).

node 184 at (0, 3), is fixed in the y-direction. These boundary conditions prevent rigid body rotation and translation. Element 1 is at the lower left-hand corner of the beam. Elements 61, 121, 181, 241, and 301 form the column of elements above element 1 at the left-hand side of the beam. Your numbering of elements and nodes may be different, but the loading and displacement boundary conditions should be the same.

2. Use your FEM program to analyze this problem using 4CST quadrilaterals, and then compare the numerical results with theory at the centroid of the upper left-hand corner element for bending stress and at node 427 where the point load is applied just beneath the point load for beam deflection. The element centroid is at $x = 0.5$ and $y = 5.5$ in.; the tip is at $x = 60$, $y = 6$ in. When iterative equation solving is used, apply 100,000 iterations (a relatively large number). Repeat the analysis using elimination equation solving. Compute and then tabulate relative errors on maximum bending stress and tip deflection. The element for maximum bending stress is the upper left-hand element with centroid at (0.5, 5.5).

   Note: Finite element analyses are almost always based on meshes that are done with the aid of a mesh generator. A mesh generator was used to prepare the meshes in Figure 1. This mesh generator has options for 4CST quadrilaterals, quadrilaterals containing four triangles, and quadrilaterals containing two triangles. The two-triangle mesh has all diagonals sloping in the same direction as seen in Figure 2c. A mesh generator with such capability is assumed to be available for this problem.

   Results should compare favorably with those in Table 1 below.

Table 1 GEOFEM 5 quadrilateral (4CST) beam results

Approach object	Elimination	Iteration (100,000)	Theory
Bending stress σ	3,649 psi	3,784 psi	3,967 psi
Relative error	8.0%	3.8%	–
Tip deflection δ	0.1799 in.	0.1850 in.	0.1920 in.
Relative error	6.3%	3.6%	–

3. Re-analyze this problem using a mesh that is all triangles, that is, four triangles per quadrilateral. Select the triangle option, but say no to the two-triangle. Tabulate results in a table similar to Table 1.
4. Re-analyze this problem using a mesh that is all triangles; only use two triangles per quadrilateral. Present the results in a table similar to Table 1.

Prepare your results in an informal report form beginning with the usual problem statement, given data, and then proceeding to a description of the procedures used and to a discussion of results in the three tables and a brief conclusion.

GEOFEM 6

This assignment explores the FEM for stress analyses in two dimensions. The problem is to compute the stress, strain, and displacement fields about a circular hole excavated in a homogeneous, isotropic, linearly elastic material. This problem has an exact mathematical solution that is obtained from elasticity theory and is often referred to as the Kirsch solution. An important lesson is that stress concentration results from FEM analyses need to be considered with due regard to element size at the hole boundary and that "small" elements are needed at the hole wall where stress changes rapidly with distance from the wall. The reason small elements are needed is to obtain reasonable design guidance, say, to evaluation of safety of a proposed underground opening, e.g., a circular tunnel or shaft.

The solution to the problem of stress distribution about a circular hole in an infinite medium is given in many texts and references books, for example "Rock Mechanics and the Design of Structures in Rock", L. Obert and W. I. Duvall, N.Y., John Wiley and Sons, Inc., 1967. Both authors were pioneers in rock mechanics and spent many years in research at the Denver Center of the former U.S. Bureau of Mines. Evidently, the first publication was by G. Kirsch (1898). With reference to the notation in Figure 1, the stresses about a circular hole in a homogenous, isotropic, linearly elastic medium are:

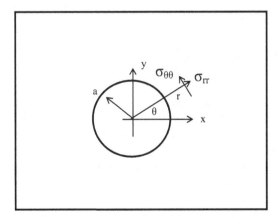

Figure 1 Notation for a circular hole in an infinite medium.

$$\sigma_{rr} = (1/2)\,(S_x + S_y)\,(1 - a^2/r^2) + (1/2)\,(S_x - S_y)\,(1 - 4a^2/r^2 + 3a^4/r^4)\cos(2\theta)$$
$$\sigma_{\theta\theta} = (1/2)\,(S_x + S_y)\,(1 + a^2/r^2) - (1/2)\,(S_x - S_y)\,(1 + 3a^4/r^4)\cos(2\theta) \tag{1}$$
$$\tau_{r\theta} = (-1/2)\,(S_x - S_y)\,(1 + 2a^2/r^2 - 3a^4/r^4)\sin(2\theta)$$

Under plane strain conditions, the z-direction shear stresses and strains are nil. The z-direction normal stress may be computed through Hooke's law, i.e., $\sigma_{zz} = v(\sigma_{rr} + \sigma_{\theta\theta})$. In Equations (1), the prehole stresses are the x and y normal stresses S_x and S_y; the other prehole stresses are nil. Note that at the hole wall, $r = a$ and the radial normal stress and shear stress are nil.

If the hole is pressurized (water or compressed air?), then Equations (1) must be changed to reflect this boundary condition. The change is simple; just add $-pa^2/r^2$ to the first equation (radial stress) in (1), recalling that pressure is compressive and therefore negative, and subtract pa^2/r^2 from the second equation in (1). Thus, an internal pressure induces a circumferential tension of equal magnitude at the hole wall (in addition to the stress associated with prehole stress).

The induced radial and circumferential displacements u and v are given by the rather long formulas:

$$u = \left(\frac{1-v^2}{E}\right)\left[\left(\frac{S_x + S_y}{2}\right)\left(r + \frac{a^2}{r}\right) + \left(\frac{S_x - S_y}{2}\right)\left(r + \frac{4a^2}{r} - \frac{a^4}{r^3}\right)\cos(2\theta)\right]$$
$$\quad - \left(\frac{v(1+v)}{E}\right)\left[\left(\frac{S_x + S_y}{2}\right)\left(r - \frac{a^2}{r}\right) - \left(\frac{S_x - S_y}{2}\right)\left(r - \frac{a^4}{r^3}\right)\cos(2\theta)\right]$$
$$v = \left(\frac{1-v^2}{E}\right)\left[-\left(\frac{S_x - S_y}{2}\right)\left(r + \frac{2a^2}{r} + \frac{a^4}{r^3}\right)\sin(2\theta)\right] \tag{2}$$
$$\quad - \left(\frac{v(1+v)}{E}\right)\left[\left(\frac{S_x - S_y}{2}\right)\left(r - \frac{2a^2}{r} + \frac{a^4}{r^3}\right)\sin(2\theta)\right]$$

At the hole wall, $r = a$ and

$$u = \left(\frac{1-v^2}{E}\right)\left[\left(\frac{S_x + S_y}{2}\right)(2a) + \left(\frac{S_x - S_y}{2}\right)(4a)\cos(2\theta)\right]$$
$$v = \left(\frac{1-v^2}{E}\right)\left[-\left(\frac{S_x - S_y}{2}\right)(4a)\sin(2\theta)\right] \tag{3}$$

where an outward radial displacement is positive and counter-clockwise circumferential displacement is also positive. Note: Tensile stresses are also positive and Equations (2) and (3) are for plane strain conditions. Equations (2) and (3) require modification if the hole is pressurized.

An interesting feature of solutions to plane problems in the mathematical theory of elasticity is the absence of elastic properties in the equations for stress. Thus, Equations (1) are valid for any homogeneous, isotropic, linearly elastic material. However, elastic properties do appear in the equations for displacement as seen in Equations (2) and (3).

A test for the validity of the stress and displacement solutions would be to differentiate the displacements to arrive at expressions for the associated strains, and then to substitute the strains into Hooke's law to obtain expressions for the stresses. Differentiation of the stresses and substitution into the equations of equilibrium should then satisfy the stress equations of equilibrium. The stresses should also satisfy the stress boundary conditions at the inner boundary (hole wall) and the far boundary at infinity. For completeness, one notes that the z-direction displacement is nil (plane strain); no variation in the z-direction occurs. For this reason, the z-direction strains are nil and the z-direction shear stresses are nil, but not the z-direction normal stress (which is?).

1. Generate the files necessary for analysis of a circular hole excavated in an initially stressed, homogeneous, isotropic elastic rock mass using a mesh generator. Specify 18 elements per ring and 20 rings for a total of 360 elements. The mesh is shown in Figure 2 below. Recall that a circle is a special ellipse (aspect ratio of 1) should your mesh generator have an ellipse option.

    The origin of coordinates is at the lower left in the standard position. Quarter symmetry is implied with the boundary conditions imposed that prevent displacement normal to the sides of the mesh. Reference to the coordinates file shows the excavated hole has a radius of 3 to the interface between gray and white elements; width (and height) of the mesh is 30. The gray ring is a ring of small elements that are excavated (cut elements). In this regard, excavation changes cut elements to "air" elements of zero stress and then computes the out-of-equilibrium forces at the boundary of the air and rock elements. These forces are cut or excavation

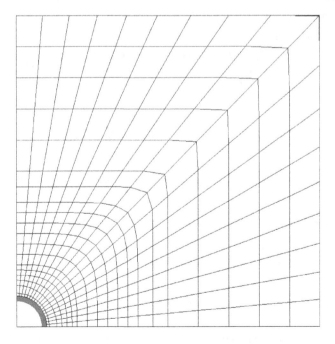

Figure 2 A plot of the circular hole mesh.

forces that are equal, but opposite in sense to the forces associated with preex-cavation (initial) stresses; they load the adjacent rock and induce displacements, strains, and stress changes in the mesh.

Notes:

(1) Use an initial stress state of $\sigma_{yy} = -1,000$ psi and all other stresses equal to zero. Use the sign convention that tension is positive.
(2) Use a Young's modulus of 2.4 (10^6) psi and a Poisson's ratio of 0.20.
(3) Use boundary conditions that prevent displacement perpendicular to the sides of the mesh at the mesh sides. Can mix, but force and displacement in the same direction are not allowed. Usually, displacement takes precedence and force is ignored when in conflict.
(4) Use either 4CST quadrilaterals which are explicitly integrated, or numerically integrated quadrilaterals.
(5) Proceed with excavation from an initial stress state as in construction practice.

2. Verify the computer program by comparing output with the known solution, that is,

(1) Plot radial and tangential normal stress concentrations as functions of angle from the horizontal x-axis using the radius to the center of the elements in the first ring of elements at the hole wall. Also plot the shear stress in the same manner.
(2) Do a single regression analysis of all stress components from all elements in the first element ring on the known values. Plot the results and show the re-gression lines in an x–y format. Use the x-axis for known values and the y-axis for finite element output. Label the plot and indicate the regression parame-ters (correlation coefficient, slope, intercept, and the number of data points). Also place the regression equation on the plot.

3. Compare 4CST results with N4Q (numerically integrated quadrilaterals) by plot-ting all the stresses from the 4CST output on the x-axis with N4Q results on the y-axis. Do a regression analysis and present the correlation coefficient and regres-sion line equation on the plot. Discuss.

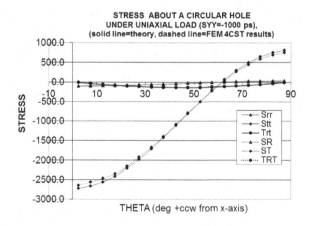

Figure 3 Stresses from theory and finite element analysis at the wall of a circular hole under uniaxial compression in the y-direction.

GEOFEM 7

Axial symmetry arises in many applications, especially in analyses of stress about drill holes and well bores under gravity loading only. In such cases, the horizontal stresses before drilling the hole, say, relative to compass coordinates, $x=$ east, $y=$ north, $z=$ up, are equal, that is, $S_{xx}=S_{yy}$. Relative to cylindrical coordinates, the prehole stress field is axially symmetric; that is, no variation with the polar angle occurs. Axial symmetry also prevails in hollow cylinders such as pipes under internal and external pressures.

In axial symmetric problems, the use of axially symmetric elements provides a substantial reduction in the number of elements needed to model the structure at hand, for example, conventional elements approximately square formed at five degree increments about a circle would number 72. Only one axially symmetric element would be needed.

Figure 1 shows a hollow cylinder under internal and external pressures.

In the elastic range of deformation, the radial, circumferential, and axial stresses are given by the formulas (*with compression positive*):

$$\sigma_r = p_b - (p_b - p_a)\left(\frac{1}{b^2} - \frac{1}{r^2}\right)\bigg/\left(\frac{1}{b^2} - \frac{1}{a^2}\right) = \sigma_3$$

$$\sigma_\theta = p_b - (p_b - p_a)\left(\frac{1}{b^2} + \frac{1}{r^2}\right)\bigg/\left(\frac{1}{b^2} - \frac{1}{a^2}\right) = \sigma_1 \qquad (1)$$

$$\sigma_z = v(\sigma_r + \sigma_\theta) = \sigma_2$$

where the ordering of principal stresses is based on the assumption that the internal pressure $p_a=0$. These equations are based on equilibrium of stresses, strain–displacement relationships, and Hooke's law for a homogeneous, isotropic, linearly elastic hollow cylinder and the conditions of the problem: axial symmetry and plane strain.

Accordingly, stress equilibrium with neglect of weight reduces to

$$\partial\sigma_r/\partial r + (\sigma_r - \sigma_\theta)/r = 0 \qquad (2)$$

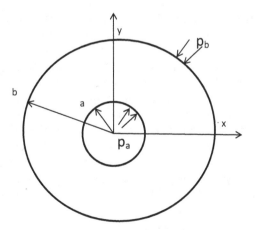

Figure 1 Diagram of a hollow cylinder under internal and external pressures.

The strain–displacement equations reduce to

$$\varepsilon_r = \partial u / \partial r, \; \varepsilon_\theta = u/r \tag{3}$$

Use of (3) in Hooke's law expresses stresses in terms of radial displacement u when substituted in (1), leading to an ordinary differential equation of second order that allows for direct integration and solution for u. The solution has the form:

$$u = C_1 r + C_2 / r \tag{4}$$

Back-substitution of (4) into (3) and then into Hooke's law gives the stresses that after imposition of boundary conditions allows for determination of the two constants of integration and finally Formulas (1). The process is straightforward, but somewhat lengthy, where careful attention to algebraic signs is essential. Thus, with pressure positive,

$$\sigma_r = p_b \left(\frac{1 - a^2/r^2}{1 - a^2/b^2} \right) = \sigma_3$$

$$\sigma_\theta = p_b \left(\frac{1 + a^2/r^2}{1 - a^2/b^2} \right) = \sigma_1 \tag{5}$$

$$\sigma_z = v(\sigma_r + \sigma_\theta) = \sigma_2$$

The questions of where the first yield occurs and under what external pressure are answered by consideration of the yield condition or failure criterion, in this case, the well-known Drucker–Prager (DP) condition. This condition in terms of principal stresses is

$$\left\{ (1/6) \left[(\sigma_1 - \sigma_2)^2 + (\sigma_2 - \sigma_3)^2 + (\sigma_3 - \sigma_1)^2 \right] \right\}^{1/2} = A(\sigma_1 + \sigma_2 + \sigma_3) + B$$

that is, $\tag{6}$

$$J_2^{1/2} = A I_1 + B, \; A = \left(1/\sqrt{3} \right)(C_0 - T_o)/(C_o + T_o), \; B = \left(2/\sqrt{3} \right)(C_o T_o)/(C_o + T_o)$$

where C_o and T_o are unconfined compressive and tensile strengths.

Of interest is the sum of the radial and tangential stresses. Inspection of (5) shows this sum to be a constant, and because the intermediate principal stress is just Poisson's ratio times this same sum, the invariant I_1 is constant. Consequently, the right-hand side of (6) is a constant through the thickness of the cylinder. The left-hand side of (6) is a lengthy expression after substitution of (5), but interestingly, all terms vary with $1/r^2$. When the left-hand side is made equal to the right-hand side, the value of r is just the value necessary to cause the first yielding. Values less than a are not admissible, of course. The first trial value of r is then a. If r is increased beyond a, the left-hand side is diminished. Thus, the first yielding must occur at $r = a$. Evaluation of (6) with $r = a$ then leads to the value of the externally applied pressure p_b^* that causes the first yielding. This value should be expressed in terms of material strengths and cylinder radii.

Further increase in p_b expands the yielding zone beyond a and ultimately to b when collapse must occur with vanishing of the elastic outer ring. The elastic–plastic solution should be found using the FEM with elastic–perfectly plastic capability and an option for the Drucker–Prager yield criterion that also serves as a plastic potential (associated flow rules).

1. Consider a hollow cylinder with an outer radius b equal to twice the inner radius a. Further suppose the cylinder ends are fixed in the axial direction, so overall a plane strain condition prevails. Apply a uniform radial pressure p_b against the outer wall of the cylinder ($r = b$).

    Use the *elastic* solution to this problem and a Drucker–Prager yield condition to determine: (1) where the first yielding occurs and (2) under what applied external pressure. Use an unconfined compressive strength of 165.5 MPa (24,000 psi) and a tensile strength of 16.5 MPa (2,400 psi) for a strength ratio of 10.0 to determine the constants A and B. Then determine the pressure p_b^* necessary to just cause yielding at the inner wall of the cylinder, $r = a$. Use a Young's modulus of 16.55 GPa (2.4 million psi) and a Poisson's ratio of 0.2. Assume isotropy.

2. Generate a mesh with a ratio of outer to inner radii of 2, and use 10 axially symmetric square elements for the finite element mesh. For example, $a = 10$ and $b = 20$. Displacement in the axial z-direction should be prevented to ensure plane strain conditions in consideration of a long hollow cylinder. The two node forces at $r = b$ are equivalent to a uniformly distributed pressure on the outer surface of the cylinder section. The total force is (pressure) (area), of course. But recall that elements are rings, so the area of pressure application A is $A = (2\pi b)(1)$. Note the element is one unit thick in the z-direction. Apply forces computed from the pressure just necessary to cause yielding at the inner wall of the cylinder.

    Compare the FEM results with the analytical results in the elastic range to the point where the first yield appears. Make the comparison graphically by plotting the radial, tangential, and axial stresses across the cylinder. Plot theoretical stress (radial, tangential, and axial) along the x-axis and FEM results along the y-axis. Then do a regression analysis of combined FEM results (all three stresses for 30 points) on theory through the cylinder.

    Repeat the analysis with a mesh of 100 square elements across a cylinder with an inner radius of 100 and an outer radius of 200. The radius ratio (b/a) remains at two, but the mesh is much more refined and thus enables a more detailed stress distribution. Also repeat the plots and regressions analysis, now with 300 points.

3. Monitoring displacements is one method of determining safety of the cylinder or well bore. Determine the relative change (percent of initial diameter) in inner diameter of the cylinder under the external pressure necessary to cause the first yielding at the well bore wall. Also determine the relative change in outer diameter at the first yield. Do this for the 10-element and 100-element meshes.

4. Discuss results and reach conclusions about axial symmetry and elastic–plastic analyses undertaken in this exercise. Be sure to address a question of mesh refinement and the question of accurate stress concentration at the limit to a purely elastic response, that is, at the first yield.

GEOFEM 8

This assignment explores nonlinear FEM stress analyses in two dimensions. The problem is to compute the stress, strain, and displacement fields about a long tunnel with a circular cross section excavated in a homogeneous, isotropic, linearly elastic material that yields in accordance with linear and quadratic forms of the well-known Drucker–Prager yield condition.

The yield condition in case of isotropy is $J_2^{n/2} + AI_1 = B$, where

> $J_2 =$ 2nd invariant of deviatoric stress,
> $I_1 =$ 1st invarinat of stress,
> $A, B =$ material constants, and
> $n =$ an exponent from testing.

In this problem, $n = 1$ and 2 (linear and quadratic forms).

Consider an initial stress state of $S_{xx} = 6.90$ MPa ($-1,000$ psi), $S_{yy} = 20.69$ MPa ($-3,000$ psi), $S_{zz} = 34.48$ MPa ($-5,000$ psi), $T_{yz} = 0$, $T_{zx} = 0$, and $T_{xy} = 0$, where x is horizontal, y is vertical, and z is along the tunnel axis (normal to the page).

This is a plane strain problem in view of the given conditions. A typical cross section and mesh are shown in the figure. This is the same mesh used in the previous elastic analyses.

Displacement normal to the sides is prevented; corners are fixed. Axes of symmetry in conjunction with sides far from the influence of the hole justify these boundary conditions.

Under plane strain conditions, the z-direction shear stresses and strains are nil. The z-direction normal stress change induced by excavation may be computed through Hooke's law, i.e., $\sigma_{zz} = v(\sigma_{rr} + \sigma_{\theta\theta})$ in the elastic domain. However, once yielding occurs and the stress–strain relations become nonlinear, a numerical calculation is required.

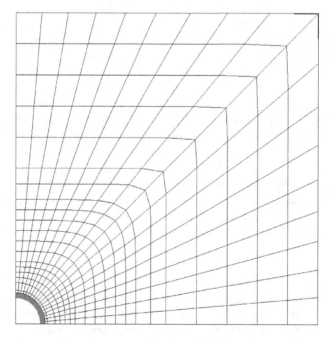

Figure 1   A plot of a circular hole mesh. Only one-quarter of the section is shown in consideration of symmetry with respect to the x- and y-axes.

Material properties for the material are as follows: Young's modulus = 17.93 GPa (2.6 million psi), Poisson's ratio = 0.26, unconfined compressive strength = 20.35 MPa (2,951 psi), tensile strength = 1.2 MPa (174 psi), and shear strength = 1.3 MPa (190 psi) in case $n = 1$. Shear strength in case $n = 2$ is 28.6 MPa (414 psi).

Use ten load increments with a maximum number of iterations of 2,000 each load step.

1. Generate the files necessary for analysis of a circular hole excavated in an initially stressed, homogeneous, isotropic elastic rock mass using a mesh generator. Specify 18 elements per ring and 20 rings for a total of 360 elements. The mesh is shown in Figure 2 below. Recall that a circle is a special ellipse (aspect ratio of 1) should your mesh generator have an ellipse option.

   The origin of coordinates is at the lower left in the standard position. Quarter symmetry is implied with the boundary conditions imposed that prevent displacement normal to the sides of the mesh. Reference to the coordinates file shows the excavated hole has a radius of 3 to the interface between gray and white elements; width (and height) of the mesh is 30. The gray ring is a ring of small elements that are excavated (cut elements). In this regard, excavation changes cut elements to "air" elements of zero stress and then computes the out-of-equilibrium forces at the boundary of the air and rock elements. These forces are cut or excavation forces that are equal, but opposite in sense to the forces associated with preexcavation (initial) stresses; they load the adjacent rock and induce displacements, strains, and stress changes in the mesh.

Figure 2 Extent of yielding (black) elements in both cases (Gauss–Seidel and conjugate gradient analyses).

Table 1 GEOFEM 7 tunnel results

Approach object	Gauss–Seidel	Conjugate gradient
Solution time (seconds)	8	4
Number of yielding elements	57	57
Stress in element near x S_{xx}, S_{yy}, S_{zz}, T_{yz}, T_{zx}, T_{xy}	−573, −7201, −6,241, 0, 0, 262 psi −3.9, −49.7, −43.0, 0, 0, 1.8 MPa	−573, −7,201, −6,241, 0, 0, 262 psi −3.9, −49.7, −43.0, 0, 0, 1.8 MPa
Stress in element near y S_{xx}, S_{yy}, S_{zz}, T_{yz}, T_{zx}, T_{xy}	294, −602, −1,245, 0, 0, 11 psi 2.0, 4.1, −8.6, 0, 0, 0.1 MPa	294, 602, −1,245, 0, 0, 11 psi 2.0, 4.1, −8.6, 0, 0, 0.1 MPa
Stress far from hole wall S_{xx}, S_{yy}, S_{zz}, T_{yz}, T_{zx}, T_{xy}	−42, −2,864, −4,976, 0, 0, 2 psi 0.3, 19.7, −34.3, 0, 0, 0 MPa	−42, −2,864, −4,976, 0, 0, 2 psi 0.3, 19.7, −34.3, 0, 0, 0 MPa

2. This exercise explores material nonlinearity in a FEM analysis using an elastic–perfectly plastic differential stress–strain law. The exploration begins with equation solving using the iterative Gauss–Seidel approach and continues with another popular iterative approach, a conjugate gradient solver. Comparisons based on extent of the yielding zone about the tunnel, accuracy, and solution time are requested, with results compared in table format as in the elastic analyses problem. Table 1 is an example of stress comparison.
3. Show the extent of yielding elements in a plot such as in Figure 2.
4. Discuss and reach conclusions from your comparison data in a short paragraph.

GEOFEM 9

This exercise concerns beam bending beyond the elastic range and invokes elastic–plastic finite element analysis considering Drucker–Prager yielding, but with the special provision of equal unconfined compressive and tensile strengths expected in case of beams used in structures such as bridges and buildings.

Use a mesh generation program to generate a beam as in the past assignment that focused on beam action. Generate a beam consisting of 12×120 square elements. In this exercise, the beam should be simply supported and loaded with a point force at mid-span. The analysis is plane strain and therefore two-dimensional with the x-axis along the beam and the y-axis vertical. In fact, the problem is one of sheet bending, but for simplicity, "beam bending" is used.

Use a Young's modulus of 16.55 GPa (2.4 (10^6) psi) and a Poisson's ratio of 0.20. Also use a Drucker–Prager yield condition in linear form ($n = 1$) with equal unconfined compressive and tensile strengths of 165.5 MPa (24,000 psi).

Use boundary conditions such that the node at the mid-height of the left-hand side of the beam (0, 6) is fixed and the corresponding node at the right-hand side (120, 6) is fixed but only in the vertical y-direction. These boundary conditions correspond to a simply supported beam with end rotation possible.

1. Do a beam analysis and plot shear and moment diagrams.
2. Determine the applied force necessary to cause the first yielding and the location of the first yielding when plastic zones just appear using technical beam bending theory. Apply a force that is 90% of the load that would just cause yielding, so the deformation is purely elastic. Then compare the finite element distribution of stress at mid-span with the theoretical solution based on technical beam bending theory. Note: This problem is slowly convergent, so when using an iterative solution, many iterations are necessary, perhaps 100,000.
3. Increase the applied force and rerun the analysis several times until sufficient data are available to plot a force–displacement curve that shows the elastic range and subsequent elastic–plastic behavior. Do the analysis using 4CST and N4Q (numerically integrated) linear displacement, constant-strain elements.
4. Determine the applied force necessary to cause collapse when the plastic zones extend through the beam section. Again use 4CST and N4Q elements.

GEOFEM 10

A useful feature of FEM is the capability of using elements of different types in the same model. For example, elements representing beams can be used with plate elements in structural models. This exercise considers a simple bar element representing a rock bolt for rock reinforcement in models of underground excavations. This element is also known as a truss bar element.

Consider a bar element in one dimension illustrated in Figure 1. The element stiffness matrix [k] may be obtained by first obtaining the element force–displacement relationship along the bar. The bar is assumed to be elastic with area A, length L, and Young's modulus E, so that

$$\sigma = E\varepsilon, \; f = \sigma A, \; d = L\varepsilon. \text{ Hence } f = (AE/L)d, \text{ so } k = AE/L \tag{1}$$

where f is the force and d is the relative displacement between ends, that is, $f = kd$ much like a linear spring.

From (1),

$$f1 = k(d1 - d2) \text{ and } f2 = k(d2 - d1) \tag{2}$$

where $(d1 - d2)$ is the displacement of the point 1 relative to 2, and $(d2 - d1)$ is the displacement of point 2 relative to 1. In matrix form, (2) is

$$\begin{Bmatrix} f1 \\ f2 \end{Bmatrix} = \begin{bmatrix} k & -k \\ -k & k \end{bmatrix} \begin{Bmatrix} d1 \\ d2 \end{Bmatrix} \text{ or } \{f\} = [k]\{d\} \tag{3}$$

f1, d1   f2, d2

Figure 1 Basic bar element.

Consider the same bar element that responds only to axial load in three space dimensions (xyz) as illustrated in Figure 2. The element is represented by the heavy arrow and may have displacements u, v, and w in the x-, y-, and z-directions. Because the element responds only to axial load, the resultant displacement of the element is either elongation or shortening of the truss bar (no bending).

If C_x, C_y, and C_z are the direction cosines of the element, then the displacements of the element ends $d1x$, $d1y$, and $d1z$ and those similarly for point 2 in the x-, y-, and z-directions are

$$d1x = C_x d1, \; d1y = C_y d1, \; d1z = C_z d1 \text{ and } d2x = C_x d2, \; d2y = C_y d2, \; d2z = C_z d2 \quad (4)$$

Similarly for the forces

$$f1x = C_x f1, \; f1y = C_y f1, \; f1z = C_z f1 \text{ and } f2x = C_x f2, \; f2y = C_y f2, \; f2z = C_z f2 \quad (5)$$

In matrix form, (4) and (5) are

$$\begin{Bmatrix} d1x \\ d1y \\ d1z \\ d2x \\ d2y \\ d2z \end{Bmatrix} = \begin{bmatrix} C_x & 0 \\ C_y & 0 \\ C_z & 0 \\ 0 & C_x \\ 0 & C_y \\ 0 & C_z \end{bmatrix} \begin{Bmatrix} d1 \\ d2 \end{Bmatrix} = [C]\{d\} \text{ and } \begin{Bmatrix} f1x \\ f1y \\ f1z \\ f2x \\ f2y \\ f2z \end{Bmatrix} = \begin{bmatrix} C_x & 0 \\ C_y & 0 \\ C_z & 0 \\ 0 & C_x \\ 0 & C_y \\ 0 & C_z \end{bmatrix} \begin{Bmatrix} f1 \\ f2 \end{Bmatrix} = [C]\{f\} \quad (6)$$

Interestingly, the products $[C]^t [C] = [C] [C]^t = [I]$, a 2×2 identity matrix. Thus,

$$\{f\}_{6\times1} = [C]_{6\times2} \{f\}_{2\times1} = [C]_{6\times2} [k]_{2\times2} [C]^t_{2\times6} \{d\}_{6\times1}, \text{ that is, } \{f\}_{6\times1} = [k]_{6\times6}\{d\} \quad (7)$$

The last equation is the truss bar element stiffness equation in three-dimensional space. This transformed element stiffness matrix is a 6×6 array where each of the two nodes have three displacements. In component form, (7) is

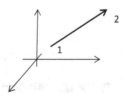

Figure 2 Bar in three dimensions.

$$\begin{Bmatrix} f1x \\ f1y \\ f1z \\ f2x \\ f2y \\ f2z \end{Bmatrix} = (k) \begin{bmatrix} C_xC_x & C_xC_y & C_xC_z & -C_xC_x & -C_xC_y & -C_xC_z \\ C_yC_x & C_yC_y & C_yC_z & -C_yC_x & -C_yC_y & -C_yC_z \\ C_zC_x & C_zC_y & C_zC_z & -C_zC_x & -C_zC_y & -C_zC_z \\ -C_xC_x & -C_xC_y & -C_xC_z & C_xC_x & C_xC_y & C_xC_z \\ -C_yC_x & -C_yC_y & -C_yC_z & C_yC_x & C_yC_y & C_yC_z \\ -C_zC_x & -C_zC_x & -C_zC_z & C_zC_x & C_zC_y & C_zC_z \end{bmatrix} \begin{Bmatrix} u1 \\ v1 \\ w1 \\ u2 \\ v2 \\ w2 \end{Bmatrix} \tag{8}$$

Consider a truss bar element parallel to the z-axis such that the direction angles from x, y, and z are 90°, 90°, and 0°, respectively. The direction cosines C_x, C_y, and C_z are then 0, 0, and 1, respectively. The only non-zero matrix elements in (8) are the C_zC_z elements. Hence,

$$f1z = k(w1 - w2) \text{ and } f2z = k(-w1 + w2) \tag{9}$$

that has the requisite form (2). If the element were aligned with the x-axis, then the direction angles would be 0°, 90°, and 90°, respectively, and C_x, C_y, and C_z would be 1, 0, and 0. According to (8) then with only C_xC_x not zero

$$f1x = k(u1 - u2) \text{ and } f2x = k(-u1 + u2) \tag{10}$$

Once forces and displacements are determined, stress and strains follow from (1).

A simplification is possible in practice by equating k to an "effective" Young's modulus E_b and thus eliminating any consideration of A and L separately. If the element were a real bar, then an effective modulus would not be appropriate. Otherwise, the element can be used to prevent overlap by adjacent surfaces. This usage occurs in simple contact algorithms that allow surfaces to move close together using a very low effective modulus value until contact is imminent when a very high modulus value is brought into play. The corresponding stress–strain plot for the truss bar would have the appearance in Figure 3. The abrupt transition in Figure 3 could be smoothed to the dotted line, which is typical of uniaxial tests on sand fill used in some mines. Such a curve may also be used to mimic compaction of caved ground formed in the course of some longwall mines. However, the curve is one-dimensional and use in finite element modeling requires a proper stress–strain relationship that links all stresses and strains for the simple reason that fill and caved material are not compacted under uniaxial load in actual practice. Indeed, the question of fill and caved material modeling goes well beyond consideration of a simple one-dimensional truss bar element.

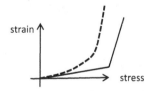

Figure 3 Stress–strain plot for overlap prevention.

Truss bar elements find use in assembled trusses illustrated in a simple manner in Figure 4, where all elements are pin-connected and thus do not support moments at the connections. The truss bar elements are thus only loaded axially. Each member of the truss can be represented by several truss bar elements, and the truss can be much more complicated in three dimensions. The truss members could be steel of various shapes, wide flange, box, and so forth. When the members are not considered pin-connected, then the elements should be beam elements that take into account bending moments as well as axial loads.

1. Consider a steel bar 16 mm (5/8 in.) in diameter and 1.83 m (6 ft) in length with a Young's modulus of 200 GPa (29 million psi) and a Poisson's ratio of 0.2. The tensile strength is 517 MPa (75,000 psi), and the ultimate strength is 688 MPa (100,000 psi). These properties correspond to a conventional rock bolt used to support rock in the vicinity of underground excavations. Note: ductility of steel suggests equal compressive and tensile strengths.

    Compute the bolt (bar) stiffness k_b. Also compute the axial stiffness k_a of a 4CST element 1.83 m (6 ft) square and the axial stiffness k_c for a N4Q element of the same dimensions. Plot force–displacement curves that show axial response in the elastic range of all three elements. Hint: Repeated single-element runs of the 4CST and N4Q elements will be necessary to gather the necessary force–displacement plots. Five points (displacement, force) for each plot in the elastic domain are suggested. These data should be tabulated before plotting.

GEOFEM II

1. Use the divergence theorem in the form of "a principle of virtual power", to derive a continuum expression for slow, incompressible fluid flow in a rigid, porous medium. Use P for fluid stress with compression negative and v for fluid displacement relative to the porous solid. Use v for fluid velocity. Do the derivation in xyz coordinates.
2. Assume a linear interpolation of pressures in a finite element, and then using conventional notation, recast the continuum result in Problem 1 into finite element form. Be sure to define all terms and matrix dimensions.
3. Verify the "Desai" example problem using a finite element seepage program (Desai and Abel).

    Note: The Darcy velocity v is a nominal velocity that when multiplied by the cross-sectional area of flow A gives the volume flow rate Q. Thus, $Q = vA$ (m³/s). The true flow velocity is faster because flow occurs only through the connected voids. The specific discharge $q = Q/A$ on average.

Figure 4 A simple truss composed of truss bar elements.

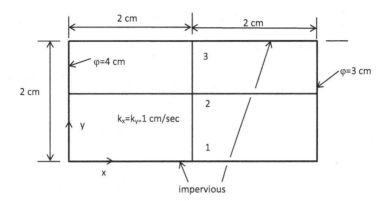

Figure 1 Mesh of four elements for a seepage analysis involving Darcy's flow. Hydraulic
conductivities are given in the figure as are piezometric heads on the left- and
right-hand sides of the figure. Top and bottom surfaces are impervious; no flow
occurs across these boundaries. Note: φ is the piezometric head.

Reference
Desai, C.S. and J.F. Abel (1972). *Introductions to the Finite Element Method: A
Numerical Method for Engineering Analysis.* Van Nostrand Reinhold Company,
New York, pp 384–389.

3. Extend the FEM analysis of seepage to accommodate compressibility of a fluid
   flowing in a compressible solid. Recall that the Darcy velocity is a velocity of the
   fluid relative to the solid velocity.

GEOFEM 12

1. Extend the FEM analysis further to take into account deformation of a saturated,
   porous, fractured solid as well as the flow of a compressible fluid in the connected
   voids of pores and cracks. This extension of FEM is a coupled hydro-mechanical
   model.
2. Program verification involves comparison of program results with known problem solutions as always. One such solution in case of deformation of saturated
   porous media is the famous Terzaghi consolidation problem. Solve a Terzaghi-like
   consolidation problem. The problem is a one-dimensional application of load to a
   column of a saturated porous material that otherwise follows Hooke's law in the
   realm of linear elasticity; fluid flow in the porous material is described by Darcy's
   law.

For the porous solid, use a Young's modulus of 68.9 MPa (10,000 psi) and a Poisson's
ratio of zero. The fluid is incompressible, so $c = 0$. Consider the column height $L = 100$.
Hydraulic conductivity is 4.88 (10^{-5}) m/s [1.602 $(10^{-4}$ ft/s)], and the specific weight of
water is 9.88 kN/m^3 (62.43 lbf/ft^3).
   An example plot of some pressure results is given in Figure 1. Be sure to find displacement as a function of time and depth as well.

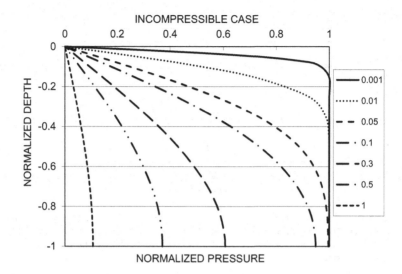

Figure 1 Pressure in the consolidating column at various times. Normalized depth is the actual depth divided by column depth. Normalized pressure is the actual pressure divided by the applied pressure. Times in the legend are dimensionless times obtained with the aid of the consolidation coefficient that is briefly discussed below in notes.

An interesting "trick" to obtain the long-term steady-state equilibrium condition is to use a large time step. Examination of the analytical solution shows this feature to be the case, provided one is cognizant of the series solution and the number to which it converges.

Notes: Units of the consolidation coefficient c_v are (L^2/T) as inspection of the diffusion equation shows. Our computation expressed $c_v = \dfrac{kE'}{1+cE'}$. The question that arises concerns the units of the "coefficient of permeability". The answer depends on how Darcy's law is written. When written using pressure gradients, the units of k are (L^4/FT). A more convenient form is the use of head gradients, so k is hydraulic conductivity with units of velocity (L/T). Hence, $c_v = \dfrac{(k/\gamma_f)E'}{1+cE'}$, which has units of

$$c_v = \frac{\left[(L/T)/(F/L^3)\right]\left(F/L^2\right)}{1+\left[1/(F/L^2)\right]\left(F/L^2\right)} = \left(L^2/T\right).$$ Units are important!

An important use of c_v is in the formation of a dimensionless time t^* such that $t^* = tc_v/L^2$ and $t = t^* L^2/c_v$. A characteristic time for consolidation to be almost complete is often approximated by a dimensionless time of one, so the actual time in a given problem is $t = L^2/c_v$. Time steps starting small at 0.0001 or so and increasing by a factor of 10 are often used in consolidation studies. Of course, the units of time in practice may be seconds, days, or years, depending on the material and the problem.

Another interesting and practical consideration besides bearing capacity of foundations occurs during blasting in wet ground. This action may be considered as applying a sudden upward force instead of a downward one in the Terzaghi consolidation. What happens to the ground about the blast or to the solid and fluid stresses in the column?

GEOFEM 13

1. Verify a hydro-mechanical couple finite element program by comparing FE results with the theoretical solution to the consolidation problem addressed previously.

    Generate a simple mesh for validation purposes, say, with a 14-element column of two-dimensional elements with nine equal square elements from the bottom ($z = +up$) and with partitioning of the tenth element at the column top into five thin elements. This refinement at the top of the mesh gives more details where force is applied.

    In this regard, there is a pressure discontinuity at the surface that usually shows as fluctuations in the first few top elements and in the Fourier series solution at small time steps, especially so if the fluid is considered incompressible. The solid is compressible in any case, or else no deformation is possible. When both are compressible, the surface undergoes an instantaneous elastic displacement as the applied total load is distributed between fluid and solid. In case of incompressible fluids and very small time steps (near zero), the applied load is carried entirely by the fluid. The pressure gradient is high at this instant, so the instantaneous fluid velocity is also very high.

    Comparison of stresses (total, effective, and fluid) via plots of data is perhaps the best way of doing code validation. However, displacements could also be compared, provided a displacement solution is computed from the fluid stress solution. Quantitative comparison of results is readily done through regression analysis. An example is shown in Figure 2.

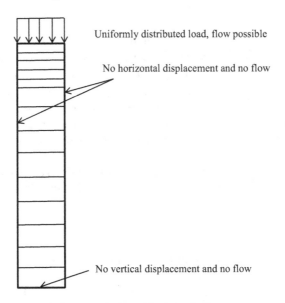

Uniformly distributed load, flow possible

No horizontal displacement and no flow

No vertical displacement and no flow

Figure 1 A 14-element mesh for coupled problem analysis.

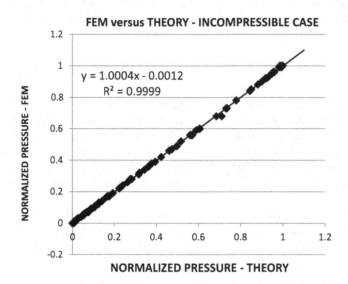

Figure 2 An example of regression analysis comparing finite element results with theory.

2. Continue with material properties used in the previous exercise. Hence, for the porous solid, use a Young's modulus of 68.9 MPa (10,000 psi) and a Poisson's ratio of zero. The fluid is incompressible, so $c = 0$. Consider the column height $L = 100$. Hydraulic conductivity is 4.88 (10^{-5}) m/s [1.602 $(10^{-4}$ ft/s)], and the specific weight of water is 9.88 kN/m³ (62.43 lbf/ft³).

    Be sure to mention the equation of solving technique and the type of element used in your finite element program and to comment on agreement or lack with theory.

Note: An important use of c_v is in the formation of a dimensionless time t^* such that $t^* = t c_v / L^2$ and $t = t^* L^2 / c_v$. A characteristic time for consolidation to be almost complete is often approximated by a dimensionless time of one, so the actual time in a given problem is $t = L^2 / c_v$. Time steps starting small at 0.0001 or so and increasing by a factor of 10 are often used in consolidation studies. Of course, the units of time in practice may be seconds, days, or years, depending on the material and the problem.

Another interesting and practical consideration besides bearing capacity of foundations occurs during blasting in wet ground. This action may be considered as applying a sudden upward force instead of a downward one in the Terzaghi consolidation. What happens to the ground about the blast or to the solid and fluid stresses in the column?

GEOFEM 14

Use a boundary element method (BEM) to compute the stress distribution about twin circular tunnels excavated in a stress field such that the vertical stress is 10 MPa and the horizontal stress is 0 MPa. Young's modulus is 10 GPa, and Poisson's ratio is 0.25. Separation of tunnels is equal to tunnel diameter.

1.   Determine the stress concentration factors at points A, B, C, and D ("Notes", Figure 14.4) and at the pillar mid-point D assuming the applied stress (preexcavation) is compressive; that is, compression is positive. Also determine the vertical stress at the center of the pillar that separates the twin tunnels. Hint: Vertical stress applies at A, B, and D; horizontal stress applies at C. Stress concentration is the actual stress divided by the major principal stress applied (10 MPa).
2.   Repeat the analysis with a horizontal loading of 10 MPa. Tabulate the results with the first analysis in a table form similar to the table following.
3.   Estimate the stress concentrations at A, B, C, and D from the table values when the preexcavation stresses are SV = 10, SH = 5 and then compute using BEM and compare as in the table. Comment.

Table of Stress Concentration at Critical Points at the Walls of Twin Circular Tunnels

Point/load	Method	A	B	C	D
SV = 10, SH = 0	BEM				
	FEM				
SV = 0, SH = 10	BEM				
	FEM				
SV = 10, SH = 5 EST	BEM				
	FEM				
SV = 10, SH = 5 COMP	BEM				
	FEM				

4.   Repeat analyses using FEM and then enter results into the table with BEM results.
5.   Briefly discuss your results and qualifications that may be appropriate.

GEOFEM 15

This exercise concerns the distinct element method (DEM) and discontinuous deformation analysis (DDA).

1.   Diagram in schematic form the first of Equations (15.2) in the text.
2.   Justify Equations (15.4) in the text with the aid of a diagram. Focus on the first of 15.4.
3.   Show that without damping, DEM block collisions will continue indefinitely. Do this demonstration by dropping a single ball or disk of specified mass from a given height. Solve the problem of energy conservation and then use a DEM program such as BALL or TRUBAL to obtain results for comparison with the exact solution.
4.   Block motion is composed of displacements associated with translation and rotation of the mass center and volume and shape changes. Show in two dimensions (x, y) that displacements (u, v) in the x- and y-directions are given by

$$u = u_o - (y - y_o)\omega + (x - x_o)\varepsilon_{xx} + (y - y_o)\gamma_{xy} \tag{1}$$

where $u, u_o, \omega, \varepsilon_{xx}$, and γ_{xy} are displacement, translation displacement, rotation angle, normal strain $(\partial u / \partial x)$, and shear strain $(1/2)(\partial u / \partial y + \partial v / \partial x)$.

Similarly for v,

$$v = v_o + (x - x_o)\omega + (y - y_o)\varepsilon_{yy} + (x - x_o)\gamma_{xy} \qquad (2)$$

with normal strain $\varepsilon_{yy} = (\partial v / \partial y)$. Note: With reference to the diagram following, for small strains, $\tan(\gamma) \doteq \gamma$.

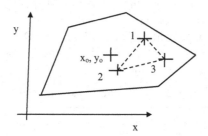

Addition of displacements gives Equations (1) and (2) for a *forward* analysis.

How large may the rotation be and can it still be considered small?

5. Show that a least square fit of displacements at three measurement points in a two-dimensional block allows for interpolation of displacements elsewhere in the block similar to finite element analysis. Consider the three measurement points in a triangular configuration as illustrated in the diagram following.

6. Consider a triangular block in two dimensions (x, y) with corners 1, 2, 3 at (2, 2), (4, 1), and (3, 3), respectively. Measured displacements (u, v) are given in the table following.

Table of displacements for Problem 6.

	u	v
1	1.515192	1.326352
2	1.658456	1.68884
3	1.326352	1.484808

Sketch the original and final positions of the block and then determine block displacements associated with translation, rotation, and strains.

$$u = a_o + a_1 x + a_2 y$$
$$v = b_o + b_1 x + b_2 y \qquad (1)$$

Appendix C

Question replies

GEOFEM I KEY

1. What is a finite "element"?

    Reply: A finite element is a homogeneous subdivision of the continuum and is quite similar to a finite difference cell or a distinct element.

2. What is the "finite element method"?

    Reply: The finite element method is a numerical technique for approximating fields for the purpose of solving boundary value problems.

3. Briefly explain the popularity of the finite element method.

    Reply: Popularity of the method stems from the ease of implementation relative to traditional finite difference methods, especially in consideration of boundary values.

4. What is meant by the "motion" of a material body?

    Reply: Motion of a material body is defined by displacements that occur in time as each material point in the body moves from one place to another. A general motion is composed of rigid body translation and rotation, volume change, and shape change. In symbolic form, a motion is

    $$x = f(X,Y,Z,t), \; y = g(X,Y,Z,t) \; z = h(X,Y,Z,t)$$

    where f, g, and h are functions of time measured from some starting time $t = 0$ and X, Y, Z are the coordinates of a mass point or "particle" at the start of the motion. A point of interest at time t is at x, y, z. Displacements u, v, w in the coordinate directions x, y, z are $u = x - X$, $v = y - Y$, $w = z - Z$. The motion is a one-to-one mapping; no holes, gaps, or overlaps are allowed, so

    $$X = f^{-1}(x, y, z, t), \; Y = g^{-1}(x, y, z, t) \; Z = h^{-1}(x, y, z, t)$$

    where the superscripts indicate inverses. If the mapping is not invertible, then fractures or gaps form, but overlap is still not allowed.

5. What are the three main categories of equations that combine to allow one to compute the motion of a body?

Reply: The three main categories of equations are the following:

1. physical laws such as the balances of linear and angular momentum,
2. kinematics or the geometry of motion, such as strain–displacement equations, and
3. material laws such as Hooke's law in linear elasticity theory.

6. Write Hooke's law in cylindrical coordinates (r, θ, z) for an isotropic material with Young's modulus E, Poisson's ratio v, and shear modulus G.
   Reply:

$$\varepsilon_{rr} = \frac{1}{E}\sigma_{rr} - \frac{v}{E}\sigma_{\theta\theta} - \frac{v}{E}\sigma_{zz}, \quad \gamma_{r\theta} = \frac{1}{G}\tau_{r\theta}$$

$$\varepsilon_{\theta\theta} = \frac{-v}{E}\sigma_{rr} + \frac{1}{E}\sigma_{\theta\theta} - \frac{v}{E}\sigma_{zz}, \quad \gamma_{\theta z} = \frac{1}{G}\tau_{\theta z}$$

$$\varepsilon_{zz} = \frac{-v}{E}\sigma_{rr} - \frac{-v}{E}\sigma_{\theta\theta} + \frac{1}{E}\sigma_{zz}, \quad \gamma_{zr} = \frac{1}{G}\tau_{zr}$$

Note: In the isotropic case, $G = \dfrac{E}{2(1+v)}$.

7. What is strain energy and how does it relate to the first law of thermodynamics (energy balance)?
   Reply: Strain energy is associated with the work done on an elastic body during slow application of external forces at constant temperature (no heat transfer; no elevation change or chemical reaction) and is stored as internal energy in relation to the first law (energy balance) of thermodynamics: $\Delta E = \Delta KE + \Delta PE + \Delta U = W + Q$. Thus, $W = \Delta E = \Delta U$, where ΔU is the change in internal energy (strain energy). Because the body is elastic, the process is reversible. Release of the applied forces allows return to the original body size and shape.

8. Consider a computer program subroutine *NMTX* (*N*-matrix) that computes interpolation functions $N_1(x, y)$, $N_2(x, y)$ for a triangular element shown in Figure 2.1 of the "Notes" such that for any unknown such as temperature T that varies linearly with position (x, y) in the triangle; that is,

$$T = a_o + a_1 x + a_2 y$$

is also given in terms of the known temperatures T_1, T_2, T_3 at the corner points (nodes) by

$$T = N_1 T_1 + N_2 T_2 + N_3 T_3$$

Derive expressions for the interpolation functions, the *N*s, and show that indeed they are linear functions of x and y.
   Reply: At the three nodes:

$$T_1 = a_o + a_1 x_1 + a_2 y_1$$
$$T_2 = a_o + a_1 x_2 + a_2 y_2$$
$$T_3 = a_o + a_1 x_3 + a_2 y_3$$

which allows for solution of the three constants a_o, a_1, a_2. Thus, by method of determinants

$$a_o = \begin{bmatrix} T_1 & x_1 & y_1 \\ T_2 & x_2 & y_2 \\ T_3 & x_3 & y_3 \end{bmatrix} \Big/ \begin{bmatrix} 1 & x_1 & y_1 \\ 1 & x_2 & y_2 \\ 1 & x_3 & y_3 \end{bmatrix},$$

$$a_1 = \begin{bmatrix} 1 & T_1 & y_1 \\ 1 & T_2 & y_2 \\ 1 & T_3 & y_3 \end{bmatrix} \Big/ \begin{bmatrix} 1 & x_1 & y_1 \\ 1 & x_2 & y_2 \\ 1 & x_3 & y_3 \end{bmatrix},$$

$$a_2 = \begin{bmatrix} 1 & x_1 & T_1 \\ 1 & x_2 & T_2 \\ 1 & x_3 & yt_3 \end{bmatrix} \Big/ \begin{bmatrix} 1 & x_1 & y_1 \\ 1 & x_2 & y_2 \\ 1 & x_3 & y_3 \end{bmatrix}$$

After reduction of the *determinants* above,

$$a_o = (T_1 p_7 + T_2 p_8 + T_3 p_9)/\Delta$$
$$a_1 = (T_1 p_1 + T_2 p_2 + T_3 p_3)/\Delta$$
$$a_o = (T_1 p_4 + T_2 p_5 + T_3 p_6)/\Delta$$

where the algebraic details for the ps are omitted, but may be found in the "Notes". After back-substitution into the original equation,

$$T = T_1(p_1 x + p_4 y + p_7)/\Delta + T_2(p_2 x + p_5 y + p_8)/\Delta + T_3(p_3 x + p_6 y + p_9)/\Delta$$

that shows

$$N_1 = (p_1 x + p_4 y + p_7)/\Delta, \ N_2 = (p_2 x + p_5 y + p_8)/\Delta, \ N_3(p_3 x + p_6 y + p_9)/\Delta.$$

Hence, $T = T_1 N_1 + T_2 N_2 + T_3 N_3$. The Ns are clearly linear functions of x and y.
9. Show that the determinant of the coefficients Δ is just twice the area A of a considered triangle.

   Reply: With reference to the triangle in the figure below, the area of the inner triangle is the area of the rectangle minus the areas of the outer triangles.

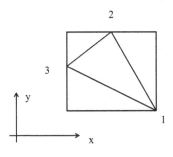

Box area $= A_a = (x_1 - x_3)(y_2 - y_3)$
Triangles: $A_{12} = (1/2)(x_1 - x_2)(y_2 - y_1)$
$A_{23} = (1/2)(x_2 - x_3)(y_2 - y_3)$
$A_{31} = (1/2)(x_1 - x_3)(y_3 - y_1)$
Area of inner triangle $= A = A_a - (A_{12} + A_{23} + A_{31})$

After multiplying and collecting terms in the last equation, the result is

$$2A = [(x_2 y_3 - x_3 y_2) + (-x_1 y_3 + x_3 y_1) + (x_1 y_2 - x_2 y_1)]$$

But this last expression is just

$$\Delta = [(x_2 y_3 - x_3 y_2) + (-x_1 y_3 + x_3 y_1) + (x_1 y_2 - x_2 y_1)] = 2A$$

So the determinant of the coefficients is equal to twice the area of the considered triangle.

10. Show that the average of an interpolated node quantity such as temperature, pressure, or displacement component at the center (x_c, y_c) of a linear triangle is just the arithmetic average of the node values. Thus,

$$T(x_c, y_c) = (1/3)(T_1 + T_2 + T_3) = (1/A)\int_A T(x, y)dA$$

where A is the area of the triangle which is actually a triangular slab one unit thick and $dV = (1)dA$, $V =$ volume. Note: The center coordinates are given by

$$x_c = (1/A)\int_A x\, dA \text{ and } y_c = (1/A)\int_A y\, dA$$

Reply:

$$T_c = (1/A)\int_A T(x, y)dA$$

$$= (1/A)\int_A (T_1 N_1 + T_2 N_2 + T_3 N_3)dA$$

$$= (1/A)\int_A \left[(T_1 p_1 + T_2 p_2 + T_3 p_3)x/\Delta + (T_1 p_4 + T_2 p_5 + T_3 p_6)y/\Delta \right.$$
$$\left. + (T_1 p_7 + T_2 p_8 + T_3 p_9)/\Delta\right]dA$$
$$= (T_1 p_1 + T_2 p_2 + T_3 p_3)x_c/\Delta + (T_1 p_4 + T_2 p_5 + T_3 p_6)y_c/\Delta + (T_1 p_7 + T_2 p_8 + T_3 p_9)/\Delta$$
$$T_c = (T_1 p_7 + T_2 p_8 + T_3 p_9)/\Delta$$

But also $x_c = (1/3)(x_1 + x_2 + x_3) = 0$ and $y_c = (1/3)(y_1 + y_2 + y_3)$, so $x_1 = (x_2 + x_3) = 0$ and $y_1 = (y_2 + y_3)$. Now

$$p_7 = x_2 y_3 - x_3 y_2,$$
$$p_8 = -x_1 y_3 + x_3 y_1 = -[-(x_2 + x_3)y_3] + x_3[-(y_2 + y_3)] = x_2 y_3 - x_3 y_2$$
$$p_9 = x_1 y_2 - x_2 y_1 = -(x_2 + x_3)y_2 - x_2[-(y_2 + y_3)] = x_2 y_3 - x_3 y_2$$

Thus, in center coordinates, $p_8 = p_9 = p_7$, and since $\Delta = p_7 + p_8 + p_9$, one has

$$T_c = (1/3)\,(T_1 + T_2 + T_3)$$

GEOFEM 2 KEY

1. Consider the tetrahedron as shown in the sketch.

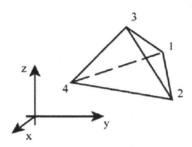

Assume a linear interpolation of displacements u, v, w in three dimensions x, y, z. Derive expressions for the interpolation functions (Ns) analogous to the two-dimensional case for a triangle. Use matrix notation for symbolic calculations, and then derive associated expressions for the six strains in three dimensions. Note: Element identification by nodes is 1, 2, 3, 4, obtained by looking down the dotted edge from node 4 to the opposing face and then listing the face nodes in counter-clockwise order. The triangular face, if it were a triangular element, would be identified by nodes 1, 2, 3 in counter-clockwise order. If the order is reversed, then during computations a sign reversal occurs that is undesirable.

Reply:

A linear interpolation for the three displacements u, v, w in the x, y, z directions is

$$\begin{aligned} u &= a_o + a_1 x + a_2 y + a_3 z \\ v &= b_o + b_1 x + b_2 y + b_3 z \\ w &= c_o + c_1 x + c_2 y + c_3 z \end{aligned} \quad \text{and for u at the nodes} \quad \begin{aligned} u_1 &= a_o + a_1 x_1 + a_2 y_1 + a_3 z_1 \\ u_2 &= a_o + a_1 x_2 + a_2 y_2 + a_3 z_2 \\ u_3 &= a_o + a_1 x_3 + a_2 y_3 + a_3 z_3 \\ u_4 &= a_o + a_1 x_4 + a_2 y_4 + a_3 z_4 \end{aligned}$$

$$= \{u_n\} = [A]\{a\}$$

The last set of equations allows for solution of the unknown constants, the as. Thus, symbolically, $\{a\} = [A]^{-1}\{u_n\}$. In determinant form,

$$a_o = \begin{vmatrix} u_1 & x_1 & y_1 & z_1 \\ u_2 & x_2 & y_2 & z_2 \\ u_3 & x_3 & y_3 & z_3 \\ u_4 & x_4 & y_4 & z_4 \end{vmatrix} \Bigg/ \begin{vmatrix} 1 & x_1 & y_1 & z_1 \\ 1 & x_2 & y_2 & z_2 \\ 1 & x_3 & y_3 & z_3 \\ 1 & x_4 & y_4 & z_4 \end{vmatrix}$$

$$= (u_1 p_1 + u_2 p_2 + u_3 p_3 + u_4 p_4)/\Delta$$

and similarly for the remaining as. Thus,

$$a_o = (u_1 p_1 + u_2 p_2 + u_3 p_3 + u_4 p_4)/\Delta$$
$$a_1 = (u_1 p_5 + u_2 p_6 + u_3 p_7 + u_4 p_8)/\Delta$$
$$a_2 = (u_1 p_9 + u_2 p_{10} + u_3 p_{11} + u_4 p_{12})/\Delta ,$$
$$a_3 = (u_1 p_{13} + u_2 p_{14} + u_3 p_{15} + u_4 p_{16})/\Delta$$

where the ps are sub-determinants in solution for the as.

The symbol Δ is the determinant of the coefficients as usual.

Alternatively, $\{a\} = [A]^{-1}\{u_n\}$. Displacement component u is now

$$u = u_1(p_1 + p_5 x + p_9 y + p_{13} z)/\Delta$$
$$+ u_2(p_2 + p_6 x + p_{10} y + p_{14} z)/\Delta$$
$$= N_1 u_1 + N_2 u_2 + N_3 u_3 + N_4 u_4$$
$$+ u_3(p_3 + p_7 x + p_{11} y + p_{15} z)/\Delta$$
$$+ u_4(p_4 + p_8 x + p_{12} y + p_{16} z)/\Delta$$

and similarly for v and w.

Alternatively, $(u)_{1x1} = \{1\,x\,y\,z\}_{1x4}[A]^{-1}_{4x4}\{u_n\}_{4x1} = [N_1\,N_2\,N_3\,N_4]\{u_n\}$.

The six strains are by definition (infinitesimal or small strain)

$$\varepsilon_{xx} = \partial u/\partial x, \quad \gamma_{xy} = \partial u/\partial y + \partial v/\partial x$$
$$\varepsilon_{yy} = \partial v/\partial y, \quad \gamma_{yz} = \partial v/\partial z + \partial w/\partial y$$
$$\varepsilon_{zz} = \partial w/\partial z, \quad \gamma_{zx} = \partial w/\partial x + \partial u/\partial z$$

The derivatives are

$$\partial u/\partial x = u_1 p_5 + u_2 p_6 + u_3 p_7 + u_4 p_8,$$
$$\partial u/\partial y = u_1 p_9 + u_2 p_{10} + u_3 p_{11} + u_4 p_{12},$$
$$\partial u/\partial z = u_1 p_{13} + u_2 p_{14} + u_3 p_{15} + u_4 p_{16}$$

$$\partial v/\partial x = v_1 p_5 + v_2 p_6 + v_3 p_7 + v_4 p_8,$$
$$\partial v/\partial y = v_1 p_9 + v_2 p_{10} + v_3 p_{11} + v_4 p_{12},$$
$$\partial v/\partial z = v_1 p_{13} + v_2 p_{14} + v_3 p_{15} + v_4 p_{16}$$

$$\partial w/\partial x = w_1 p_5 + w_2 p_6 + w_3 p_7 + w_4 p_8,$$
$$\partial w/\partial y = w_1 p_9 + w_2 p_{10} + w_3 p_{11} + w_4 p_{12},$$
$$\partial w/\partial z = w_1 p_{13} + w_2 p_{14} + w_3 p_{15} + w_4 p_{16},$$

Thus, from the above and definitions,

$$\varepsilon_{xx} = u_1 p_5 + u_2 p_6 + u_3 p_7 + u_4 p_8,$$
$$\gamma_{xy} = u_1 p_9 + u_2 p_{10} + u_3 p_{11} + u_4 p_{12} + v_1 p_5 + v_2 p_6 + v_3 p_7 + v_4 p_8$$
$$\varepsilon_{yy} = v_1 p_9 + v_2 p_{10} + v_3 p_{11} + v_4 p_{12},$$
$$\gamma_{yz} = v_1 p_{13} + v_2 p_{14} + v_3 p_{15} + v_4 p_{16} + w_1 p_9 + w_2 p_{10} + w_3 p_{11} + w_4 p_{12},$$
$$\varepsilon_{zz} = w_1 p_{13} + w_2 p_{14} + w_3 p_{15} + w_4 p_{16},$$
$$\gamma_{zx} = w_1 p_5 + w_2 p_6 + w_3 p_7 + w_4 p_8 + u_1 p_{13} + u_2 p_{14} + u_3 p_{15} + u_4 p_{16}$$

In matrix form, these equations are

$$
\begin{Bmatrix} \varepsilon_{xx} \\ \varepsilon_{yy} \\ \varepsilon_{zz} \\ \gamma_{xy} \\ \gamma_{yz} \\ \gamma_{zx} \end{Bmatrix} =
\begin{bmatrix}
p_5 & p_6 & p_7 & p_8 & 0 & 0 & 0 & 0 & 0 & 0 & 0 & 0 \\
0 & 0 & 0 & 0 & p_9 & p_{10} & p_{11} & p_{12} & 0 & 0 & 0 & 0 \\
0 & 0 & 0 & 0 & 0 & 0 & 0 & 0 & p_{13} & p_{14} & p_{15} & p_{16} \\
p_9 & p_{10} & p_{11} & p_{12} & p_5 & p_6 & p_7 & p_8 & 0 & 0 & 0 & 0 \\
0 & 0 & 0 & 0 & p_{13} & p_{14} & p_{15} & p_{16} & p_9 & p_{10} & p_{11} & p_{12} \\
p_{13} & p_{14} & p_{15} & p_{16} & 0 & 0 & 0 & 0 & p_5 & p_6 & p_7 & p_8
\end{bmatrix}
\begin{Bmatrix} u_1 \\ u_2 \\ u_3 \\ u_4 \\ v_1 \\ v_2 \\ v_3 \\ v_4 \\ w_1 \\ w_2 \\ w_3 \\ w_4 \end{Bmatrix}
$$

that are organized by components.

A somewhat different organization can be obtained by nodes rather than by components. In either case, one has in compact matrix notation

$$\{\varepsilon\} = [B]\{\delta\}$$

which is a common notation in the technical literature. Note that the B-matrix is a matrix of constants.

2. Consider the triangle in the sketch below, then use the vector (cross) product to derive an expression for the area of the triangle. Show that the determinant of the coefficients is equal to twice the area of the triangle using notation from the course notes.

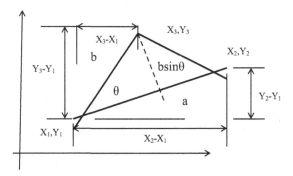

Reply:

The vector or cross product is symbolically $\mathbf{a} \times \mathbf{b}$, where bold indicates vector. In matrix form with $\mathbf{i}$, $\mathbf{j}$, $\mathbf{k}$ as unit vectors along the coordinate axes xyz, one has the determinant

$$
\begin{vmatrix}
i & j & k \\
a_x & a_y & 0 \\
b_x & b_y & 0
\end{vmatrix} = k(a_x b_y - a_y b_x)
$$

which is a vector with magnitude

$$\Delta = (x_2 - x_1)(y_3 - y_1) - (y_2 - y_1)(x_3 - x_1)$$
$$= x_2 y_3 - x_2 y_1 - x_1 y_3 + x_1 y_1 - y_2 x_3 + y_2 x_1 + y_1 x_3 - y_1 x_1$$
$$= (x_2 y_3 - y_2 x_3) + (-x_1 y_3 + x_3 y_1) + (x_1 y_2 - x_2 y_1)$$
$$\Delta = p_7 + p_8 + p_9$$

where the determinant Δ is the same as in the "Notes" as a comparison shows. In the previous problem GEOFEM 1, this determinant was shown to be just twice the area of the triangle considered.

GEOFEM 3 KEY

1. Consider a quadrilateral element shown in the sketch.

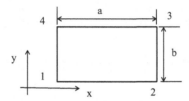

Assume a linear interpolation of temperature T in two dimensions, x and y, so that

$$T = A + Bx + Cy + Dxy$$

Shift the origin of coordinates to point 1, then derive the function for the rectangle, and bring into the form

$$T = N_1 T_1 + N_2 T_2 + N_3 T_3 + N_4 T_4$$

Note: $T(0,0) = T_1$, $T(a,0) = T_2$ $T(a,b) = T_3$, $T(0,b) = T_4$. Derive expressions for the interpolation functions (Ns), for example $N_3 = (xy)/(ab)$.

    Reply: First, find expressions for the constants: A, B, C, D. One has at point 1, $T_1 = A$; at point 2, $T_2 = T_1 + aB$ so $B = (T_2 - T_1)/a$; and so on for points 3 and 4, such that $C = (T_4 - T_1)/b$ and $D = (T_1 - T_2 + T_3 - T_4)/ab$. After back-substitution into the original equation

$$T = T_1\left(1 - \frac{x}{a} - \frac{y}{b} + \frac{xy}{ab}\right) + T_2\left(\frac{x}{a} - \frac{xy}{ab}\right) + T_3\left(\frac{xy}{ab}\right) + T_4\left(\frac{y}{b} - \frac{xy}{ab}\right). \text{ Alternatively}$$

$$T = N_1 T_1 + N_2 T_2 + N_3 T_3 + N_4 T_4$$

Thus,

$$N_1 = \left(1 - \frac{x}{a} - \frac{y}{b} + \frac{xy}{ab}\right), \quad N_2 = \left(\frac{x}{a} - \frac{xy}{ab}\right), \quad N_3 = \left(\frac{xy}{ab}\right), \quad N_4 = \left(\frac{y}{b} - \frac{xy}{ab}\right)$$

2. Suppose the quadrilateral in the sketch has node coordinates and temperatures

$T_1 = 100°, x_1 y_1 : (2,2)$ $T_2 = 60°, x_2 y_2 : (4,2)$
$T_3 = 50°, x_3 y_3 : (4,3)$ $T_4 = 90°, x_4 y_4 : (2,3)$

Using the results from Problem 1, find the temperatures at (2.5, 2.5). What is the temperature at the center of the element and what are the coordinates there?

Reply: First, relocate the origin to point 1 as in problem one, so the given data are now

$T_1 = 100°$, $x_1 y_1$: (0, 0) $T_2 = 60°$, $x_2 y_2$: (2, 0)
$T_3 = 50°$, $x_3 y_3$: (2, 1) $T_4 = 90°$, $x_4 y_4$: (0, 1)

and the given point is now at (1.0, 0.5). The equation for temperature is

$$T = (100)\left(1 - \frac{x}{2} - \frac{y}{1} + \frac{xy}{2}\right) + 60\left(\frac{x}{2} - \frac{xy}{2}\right) + 50\left(\frac{xy}{1}\right) + 90\left(\frac{y}{1} - \frac{xy}{2}\right)$$

$$= (100)\left(1 - \frac{1}{2} - \frac{0.5}{1} + \frac{0.5}{2}\right) + 60\left(\frac{1}{2} - \frac{0.5}{2}\right) + 50\left(\frac{0.5}{1}\right) + 90\left(\frac{0.5}{1} - \frac{0.5}{2}\right)$$

$$= (100)\left(\frac{1}{4}\right) + 60\left(\frac{1}{4}\right) + 50\left(\frac{1}{4}\right) + 90\left(\frac{1}{4}\right)$$

$$= (100 + 60 + 50 + 90)\left(\frac{1}{4}\right)$$

$T = 75°$

This result is also the temperature at the center of the element and is seen to be just the arithmetic average of the node temperatures.

3. Sketch the temperature distribution above the xy-plane ($T=z$ coordinate over the element in Problem 2).

   Reply:

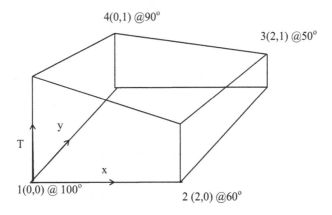

Plot for problem 3. Temperature varies linearly along the rectangle sides.

4. Determine the B-matrix for the quadrilateral shown in the sketch, so

$$\begin{Bmatrix} \varepsilon_{xx} \\ \varepsilon_{yy} \\ \gamma_{xy} \end{Bmatrix} = [B] \begin{Bmatrix} u_1 \\ \cdots \\ u_4 \\ v_1 \\ \cdots \\ v_4 \end{Bmatrix} = [B]\{\delta\}$$

Reply: 2 (2, 0)@60°

By definition, the strains are $\begin{Bmatrix} \varepsilon_{xx} \\ \varepsilon_{yy} \\ \gamma_{xy} \end{Bmatrix} = \begin{Bmatrix} \partial u/\partial x \\ \partial v/\partial y \\ \partial u/\partial y + \partial v/\partial x \end{Bmatrix}$. Hence,

$\varepsilon_{xx} = (\partial N_1/\partial x)u_1 + (\partial N_2/\partial x)u_2 + (\partial N_3/\partial x)u_3 + (\partial N_4/\partial x)u_4$. Thus,

$\varepsilon_{xx} = \left(-\dfrac{1}{a}+\dfrac{y}{ab}\right)u_1 + \left(\dfrac{1}{a}-\dfrac{y}{ab}\right)u_2 + \left(\dfrac{y}{ab}\right)u_3 + \left(-\dfrac{y}{ab}\right)u_3$ from Problem 1. Simi-

larly, $\varepsilon_{yy} = \left(-\dfrac{1}{b}+\dfrac{x}{ab}\right)v_1 + \left(-\dfrac{x}{ab}\right)v_2 + \left(\dfrac{x}{ab}\right)v_3 + \left(\dfrac{1}{b}-\dfrac{x}{ab}\right)v_3$ and for the shear strain.

$\gamma_{xy} = \left(-\dfrac{1}{b}+\dfrac{x}{ab}\right)u_1 + \left(-\dfrac{x}{ab}\right)u_2 + \left(\dfrac{x}{ab}\right)u_3 + \left(\dfrac{1}{b}-\dfrac{x}{ab}\right)u_3$

$+ \left(-\dfrac{1}{a}+\dfrac{y}{ab}\right)v_1 + \left(\dfrac{1}{a}-\dfrac{y}{ab}\right)v_2 + \left(\dfrac{y}{ab}\right)v_3 + \left(-\dfrac{y}{ab}\right)v_3$

The B-matrix for the quadrilateral in the sketch is therefore

$$[B]=\begin{bmatrix} \left(-\dfrac{1}{a}+\dfrac{y}{ab}\right) & \left(\dfrac{1}{a}-\dfrac{y}{ab}\right) & \left(\dfrac{y}{ab}\right) & \left(-\dfrac{y}{ab}\right) & 0 & 0 & 0 & 0 \\ 0 & 0 & 0 & 0 & \left(-\dfrac{1}{b}+\dfrac{x}{ab}\right) & \left(-\dfrac{x}{ab}\right) & \left(\dfrac{x}{ab}\right) & \left(\dfrac{1}{b}-\dfrac{x}{ab}\right) \\ \left(-\dfrac{1}{b}+\dfrac{x}{ab}\right) & \left(-\dfrac{x}{ab}\right) & \left(\dfrac{x}{ab}\right) & \left(\dfrac{1}{b}-\dfrac{x}{ab}\right) & \left(-\dfrac{1}{a}+\dfrac{y}{ab}\right) & \left(\dfrac{1}{a}-\dfrac{y}{ab}\right) & \left(\dfrac{y}{ab}\right) & \left(-\dfrac{y}{ab}\right) \end{bmatrix}$$

GEOFEM 4 KEY

1. Consider a generic quadrilateral element shown in the sketch.

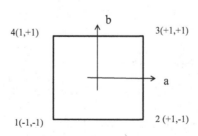

Consider the origin of coordinates at the center. Assume a linear interpolation, for example

$$u = \alpha_o + \alpha_1 a + \alpha_2 b + \alpha_3 ab \qquad (1)$$

where u is the b-component of displacement. Show algebraically that (1) can be brought into the form

$$u = N_1 u_1 + N_2 u_2 + N_3 u_3 + N_4 u_4 \qquad (2)$$

Reply:
From (1),

$$\begin{aligned}
u_1 &= \alpha_o - \alpha_1 - \alpha_2 + \alpha_3 \\
u_2 &= \alpha_o + \alpha_1 - \alpha_2 - \alpha_3 \\
u_3 &= \alpha_o + \alpha_1 + \alpha_2 + \alpha_3 \\
u_4 &= \alpha_o - \alpha_1 + \alpha_2 - \alpha_3
\end{aligned} \qquad (3)$$

Hence, $\alpha_o = (1/4)(u_1 + u_2 + u_3 + u_4)$ by summing (3). Addition of the first two of (3) and solving gives $\alpha_2 = (1/4)(u_3 + u_4) - (1/4)(u_1 + u_2)$. Proceeding in the same manner gives

$\alpha_3 = (1/4)(u_1 + u_4) - (1/4)(u_2 + u_4)$ and $\alpha_1 = (1/4)(u_2 + u_3) - (1/4)(u_1 + u_4)$. Substitution of these results into (2) gives

$$\begin{aligned}
u =\ &(1/4)(u_1 + u_2 + u_3 + u_4) + (a/4)\big[(1/4)(u_2 + u_3) - (1/4)(u_1 + u_4)\big] \\
&+ (b/4)\big[(1/4)(u_3 + u_4) - (1/4)(u_1 + u_2)\big] + (ab/4)\big[(1/4)(u_1 + u_3) - (1/4)(u_2 + u_4)\big]
\end{aligned}$$

After collecting terms

$$4u = u_1(1 - a - b + ab) + u_2(1 + a - b - ab) + u_3(1 + a + b + ab) + u_4(1 - a + b - ab)$$

2. Derive expressions for the interpolation functions in (2) such that

$$\begin{aligned}
N_1 &= (1/4)(1 - a)(1 - b), \quad N_2 = (1/4)(1 + a)(1 - b), \\
N_3 &= (1/4)(1 + a)(1 + b), \quad N_4 = (1/4)(1 - a)(1 + b)
\end{aligned}$$

Reply:
From Problem 1,

$$u = (u_1/4)(1 - a)(1 - b) + (u_2/4)(1 + a)(1 - b) + (u_3/4)(1 + a)(1 + b) + (u_4/4)(1 - a)(1 + b)$$

that can be brought into the form

$$u = (u_1/4)(1 - a)(1 - b) + (u_2/4)(1 + a)(1 - b) + (u_3/4)(1 + a)(1 + b) + (u_4/4)(1 - a)(1 + b).$$

Thus, $\begin{aligned} N_1 &= (1/4)(1 - a)(1 - b), \quad N_2 = (1/4)(1 + a)(1 - b), \\ N_3 &= (1/4)(1 + a)(1 + b), \quad N_4 = (1/4)(1 - a)(1 + b) \end{aligned}$ which are the interpolation functions. A similar development holds for v.

3. Consider the isoparametric quadrilateral shown in the sketch below, where $x = S_1 x_1 + S_2 x_2 + S_3 x_3 + S_4 x_4$ and similarly for y.

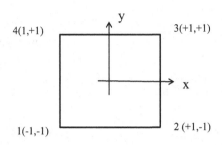

Show that

$$S_1 = (1/4)(1-x)(1-y), \quad S_2 = (1/4)(1+x)(1-y),$$
$$S_3 = (1/4)(1+x)(1+y), \quad S_4 = (1/4)(1-x)(1+y)$$

and hence

$$x = (x_1/4)(1-x)(1-y)+(x_2/4)(1+x)(1-y)+(x_3/4)(1+x)(1+y)+(x_4/4)(1-x)(1+y)$$

and similarly for y. Verify the center is at (0, 0).

Reply:

The solution follows the same path as in Problems 1 and 2 with the substitutions of x for a and y for b. Thus,

$$x = (x_1/4)(1-x)(1-y)+(x_2/4)(1+x)(1-y)+(x_3/4)(1+x)(1+y)+(x_4/4)(1-x)(1+y)$$

and similarly for y, that is,

$$y = (y_1/4)(1-x)(1-y)+(y_2/4)(1+x)(1-y)+(y_3/4)(1+x)(1+y)+(y_4/4)(1-x)(1+y)$$

Substituting (0, 0) into the expressions for x and y to obtain (x_c, y_c) shows that

$$x_c = (x_1/4)+(x_2/4)+(x_3/4)+(x_4/4)$$
$$= (-1_1/4)+(+1/4)+(+1/4)+(-1/4)$$
$$x_c = 0$$

and similarly for y. Indeed, the center (x_c, y_c) is at (0, 0).

4. Consider a one-dimensional truss bar element shown in the sketch. Assume a linear interpolation of displacement u along the bar in the x-direction such that $u = a_o + a_1 x$. Derive the Ns so that $u = N_1 u_1 + N_2 u_2$, and then find a B-matrix that gives strain in terms of the node displacements (δ_1, δ_2) at the left- and right-hand sides of the bar. Finally, assuming the bar area is A and that tensile forces are applied at the bar ends, determine the bar stiffness matrix in the linearly elastic case, i.e., find $\{f\} = [k]\{\delta\}$ in this one-dimensional case.

Hint: Start with Hooke's law for this last part.

Reply:

Consider a truss bar element in one dimension as illustrated in Figure 1. The element stiffness matrix [k] may be obtained by first finding the element force–displacement relationship along the bar. The bar is assumed to be elastic and to have an area A, length L, and Young's modulus E, so that

$\sigma = E\varepsilon$, $f = \sigma A$, $\delta = L\varepsilon$. Hence $f = (AE/L)\delta$, so $k = AE/L$

where f is the force and d is the relative displacement between ends, that is, $\{f\} = [k]\{\delta\}$.

$$f_1, \delta_1 \qquad\qquad f_2, \delta_2$$

Figure 1 Basic truss bar element.

From the above relationships,

$f_1 = k(\delta_1 - \delta_2)$ and $f_2 = k(\delta_2 - \delta_1)$

where $(\delta_1 - \delta_2)$ is the displacement of the point 1 relative to 2, and $(\delta_2 - \delta_1)$ is the displacement of point 2 relative to 1. In matrix form, (2) is

$$\begin{Bmatrix} f1 \\ f2 \end{Bmatrix} = \begin{bmatrix} k & -k \\ -k & k \end{bmatrix} \begin{Bmatrix} \delta_1 \\ \delta_2 \end{Bmatrix} \text{ or } \{f\} = [k]\{\delta\}$$

GEOFEM 5 KEY

Code validation is an essential step toward development of a reliable numerical analysis technique. The simplest validation step is to compare solutions to known problems with results from a new or unfamiliar finite element computer program. As an example, a solution to the problem of a tip-loaded cantilever beam with finite element method (FEM) results is desired. The beam bending solution may be achieved using technical beam bending theory, that is, the flexure formula and the Euler–Bernoulli bending equations. Both may be found in mechanics of materials textbooks. This comparison is just the first step in a series of such comparisons.

In addition to code validation, this exercise also concerns behavior of constant-strain triangles (CSTs) and quadrilaterals composed of four constant-strain triangles (4CSTs) under bending stress. Also of interest is a comparison of Gauss elimination equation solving with Gauss–Seidel iteration.

A cantilever beam is shown in the Figure that has a span of 60 in., a depth of 6 in., and a breadth of 1 in. into the page. The span depth ratio is 10, so beam action is assured. A point load is applied at the upper right-hand corner, while the left-hand side of the beam is fixed.

1. Consider an elastic analysis of this beam using the flexure and bending formulas that can be found in any mechanics of materials textbook. In particular, derive formulas for the bending stress (x-direction normal stress) as a function of position in the beam and for the deflection (y-displacement). Use a Young's modulus $E = 9.6 \times 106$ psi and a Poisson's ratio $v = 0.20$. Because the FEM approach is plane strain, analysis of this beam should also be plane strain. Thus, in place of E in the usual (plane stress) beam formulas, one should use $E/(1 - v^2)$.

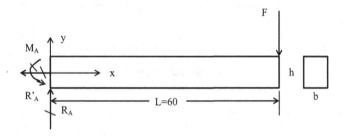

Figure I GEOFEM 5 (tip-loaded cantilever beam).

Reply:
Bending formula from technical bending theory:

$\sigma = -My/I$, M = bending moment,

I = area moment of inertial about the neutral axis,

$y = +$up, tension$+$

$d^2v/dx^2 = M/EI$, v = displacement, $+$up,

E = Young's modulus, use $E' = E/(1 - v^2)$ for plane strain

Free body diagram (overall):

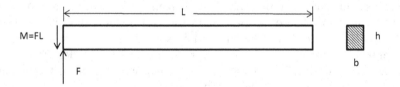

Internal equilibrium:

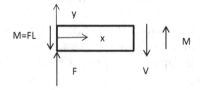

$\sum_y \text{forces} = 0, \text{therefore}, V = F \text{ for all } x$

$\sum_{x=0} \text{moments} = 0, \text{ therefore, } M = Vx - FL \text{ for all } x, \text{ i.e., } M = F(x - L), \text{ so}$

$\sigma = -yF(x - L)/I$, where $I = bh^3/12$: ("bending stress").

Integration for slope and displacement:

$d^2v/dx^2 = M/E'I$, hence
$d^2v/dx^2 = F(x - L)/E'I$, and
$\quad dv/dx = (1/E'I)(F)(x^2/2 - Lx) + C_1$, at $x = 0$, the slope is zero,
$\qquad$ so the constant is zero. Thus,
$\qquad v = (1/E'I)(F)(x^3/6 - Lx^2/2) + C_2$, at $x = 0$, the displacement
$\qquad$ is zero, so the constant is zero and
$\qquad v = (1/E'I)(F)(x^3/6 - Lx^2/2)$: ("deflection")

At (0.5, 2.5) relative to the neutral axis,

$\sigma = -(2.5)(480)(0.5 - 50)/18$, where $I = (1)(6)^3/12$
$\sigma = 3{,}967$ psi (tension+).

The deflection under the load point is

$v = \left\{1 / \left[(9.6)(10^6)/(1 - 0.2^2)18 \right] \right\}(480)\left(60^3/6 - 60(60^2)/2\right)$: ("deflection")
$v = -0.1920$ in. (negative v is down)

Three finite element meshes of the beam in Figure 1 are shown in Figure 2. There are 360 elements ($1 \times 1 \times 1$ in.) and 427 nodes in the quadrilateral mesh. There are 60 horizontal subdivisions and 6 vertical subdivisions in Figure 2a. Thus, the elements are unit squares in Figure 2a. Origin of coordinates is at the lower left-hand corner. Node number 1 is at the origin; numbering proceeds from left to right starting at the bottom and proceeding to the top. Elements are numbered in the same way.

$\qquad$ The point load is applied at node 427, the upper right-hand corner in Figure 2a. Nodes 1, 62, 123, 184, 245, 306, and 367 are at the left-hand side of the beam and are fixed in the x-direction. Additionally, the center node at the left-hand side, node 184 at (0, 3), is fixed in the y-direction. These boundary conditions prevent rigid body rotation and translation. Element 1 is at the lower left-hand corner of the beam. Elements 61, 121, 181, 241, and 301 form the column of elements above element 1 at the left-hand side of the beam. Your numbering of elements and nodes may be different, but the loading and displacement boundary conditions should be the same.

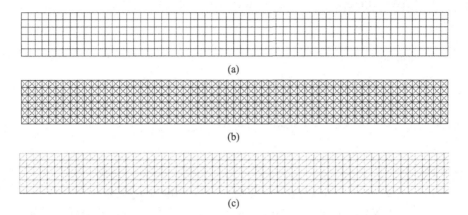

(a)

(b)

(c)

Figure 2 GEOFEM 5: (a) quadrilaterals – 4CSTs; (b) triangles (4); (c) triangles (2).

2. Use your FEM program to analyze this problem using 4CST quadrilaterals, and then compare the numerical results with theory at the centroid of the upper left-hand corner element for bending stress and at node 427 where the point load is applied just beneath the point load for beam deflection. The element centroid is at $x = 0.5$ and $y = 5.5$ in.; the tip is at $x = 60$, $y = 6$ in. When iterative equation solving is used, apply 100,000 iterations (a relatively large number). Repeat the analysis using elimination equation solving. Compute and then tabulate relative errors on maximum bending stress and tip deflection. The element for maximum bending stress is the upper left-hand element with centroid at (0.5, 5.5).

Note: Finite element analyses are almost always based on meshes that are done with the aid of a mesh generator. A mesh generator was used to prepare the meshes in Figure 1. This mesh generator has options for 4CST quadrilaterals, quadrilaterals containing four triangles, and quadrilaterals containing two triangles. The two-triangle mesh has all diagonals sloping in the same direction as seen in Figure 2c. A mesh generator with such capability is assumed to be available for this problem.

Results should compare favorably with those in Table 1 below.

Table 1 GEOFEM 5 quadrilateral (4CST) beam results

Approach object	Elimination	Iteration (100,000)	Theory
Bending stress σ	3,649 psi	3,784 psi	3,967 psi
Relative error	8.0%	3.8%	–
Tip deflection δ	−0.1799	−0.1850 in.	−0.1920 in.
Relative error	6.3%	3.6%	–

Note: Relative error is the percentage: (100) (Theory − FEM)/(Theory).

3. Re-analyze this problem using a mesh that is all triangles, that is, four triangles per quadrilateral. Select the triangle option, but say no to the two-triangle. Tabulate results in a table similar to Table 1.

Reply:

Table 2 GEOFEM 5 triangles (4) beam results

Approach object	Elimination	Iteration (100,000)	Theory
Bending stress σ	2,485 psi	2,692 psi	3,967 psi
Relative error	37%	32%	–
Tip deflection δ	−0.1322	−0.1388 in.	−0.1920 in.
Relative error	31%	28%	–

4. Re-analyze this problem using a mesh that is all triangles; only use two triangles per quadrilateral. Present the results in a table similar to Table 1.
   Reply:

Table 3 GEOFEM 5 triangles (2) beam results

Approach object	Elimination	Iteration (100,000)	Theory
Bending stress σ	3,056 psi	3,382 psi	3,967 psi
Relative error	23%	15%	–
Tip deflection δ	0.1563	0.1666 in.	0.1920 in.
Relative error	19%	13%	–

Prepare your results in an informal report form beginning with the usual problem statement, given data, and then proceeding to a description of the procedures used and to a discussion of results in the three tables and a brief conclusion.
   Reply:
   The problem is to analyze beam bending under a point load using the FEM in application to three meshes: one using only 4CST elements, the second using four triangles per 4CST element, and the third using two triangles per 4CST element as shown in Figure 2 of the assignment.

   The given data include the beam Young's modulus and Poisson's ratio, dimensions, point load magnitude, and location. A mesh generator and finite element program with iteration and elimination solvers are also given. The number of iterations is specified at 100,000 when using the iterative solver.

   The procedure is to generate the mesh and companion run stream as input to the finite element program and then to execute the program. Following successful execution, results are tabulated in the specified table format.

   The overall results indicate best accuracy using the 4CST element in conjunction with an iterative solver. Why iteration is more accurate than elimination is somewhat surprising; the reason is unclear at this juncture. One notes that residuals are fixed in elimination solving, while they are variable subject to the number of iterations when using the iterative solver, although machine word length and accuracy play a part. In this regard, double precision is required when a 32-bit word length is used. However, almost all PCs now are 64-bit machines. The two-triangle mesh leads to results more accurate than the four-triangle mesh. Again, iteration is more accurate than elimination.

GEOFEM 6 KEY

This assignment explores the FEM for stress analyses in two dimensions. The problem is to compute the stress, strain, and displacement fields about a circular hole excavated in a homogeneous, isotropic, linearly elastic material. This problem has an exact mathematical solution that is obtained from elasticity theory and is often referred to as the Kirsch solution. An important lesson is that stress concentration results from FEM analyses need to be considered with due regard to element size at the hole boundary and that "small" elements are needed at the hole wall where stress changes rapidly with distance from the wall. The reason small elements are needed is to obtain reasonable design guidance, say, to evaluation of safety of a proposed underground opening, e.g., a circular tunnel or shaft.

The solution to the problem of stress distribution about a circular hole in an infinite medium is given in many texts and references books, for example "Rock Mechanics and the Design of Structures in Rock", L. Obert and W. I. Duvall, N.Y., John Wiley and Sons, Inc., 1967. Both authors were pioneers in rock mechanics and spent many years in research at the Denver Center of the former U.S. Bureau of Mines. Evidently, the first publication was by G. Kirsch (1898). With reference to the notation in Figure 1, the stresses about a circular hole in a homogenous, isotropic, linearly elastic medium are:

$$\sigma_{rr} = (1/2)\,(S_x + S_y)\,(1 - a^2/r^2) + (1/2)\,(S_x - S_y)\,(1 - 4a^2/r^2 + 3a^4/r^4)\cos(2\theta)$$
$$\sigma_{\theta\theta} = (1/2)\,(S_x + S_y)\,(1 + a^2/r^2) - (1/2)\,(S_x - S_y)\,(1 + 3a^4/r^4)\cos(2\theta) \qquad (1)$$
$$\tau_{r\theta} = (-1/2)\,(S_x - S_y)\,(1 + 2a^2/r^2 - 3a^4/r^4)\sin(2\theta)$$

Under plane strain conditions, the z-direction shear stresses and strains are nil. The z-direction normal stress may be computed through Hooke's law, i.e., $\sigma_{zz} = \nu(\sigma_{rr} + \sigma_{\theta\theta})$. In Equations (1), the prehole stresses are the x and y normal stresses S_x and S_y; the other prehole stresses are nil. Note that at the hole wall, $r = a$ and the radial normal stress and shear stress are nil.

If the hole is pressurized (water or compressed air), then Equations (1) must be changed to reflect this boundary condition. The change is simple; just add $-pa^2/r^2$ to the first equation (radial stress) in (1), recalling that pressure is compressive and therefore negative, and subtract pa^2/r^2 from the second equation in (1). Thus, an internal pressure induces a circumferential tension of equal magnitude at the hole wall (in addition to the stress associated with prehole stress).

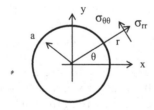

Figure 1 Notation for a circular hole in an infinite medium.

The induced radial and circumferential displacements u and v are given by the rather long formulas:

$$u = \left(\frac{1-v^2}{E}\right)\left[\left(\frac{S_x+S_y}{2}\right)\left(r+\frac{a^2}{r}\right)+\left(\frac{S_x-S_y}{2}\right)\left(r+\frac{4a^2}{r}-\frac{a^4}{r^3}\right)\cos(2\theta)\right]$$
$$-\left(\frac{v(1+v)}{E}\right)\left[\left(\frac{S_x+S_y}{2}\right)\left(r-\frac{a^2}{r}\right)-\left(\frac{S_x-S_y}{2}\right)\left(r-\frac{a^4}{r^3}\right)\cos(2\theta)\right]$$

$$v = \left(\frac{1-v^2}{E}\right)\left[-\left(\frac{S_x-S_y}{2}\right)\left(r+\frac{2a^2}{r}+\frac{a^4}{r^3}\right)\sin(2\theta)\right]$$
$$-\left(\frac{v(1+v)}{E}\right)\left[\left(\frac{S_x-S_y}{2}\right)\left(r-\frac{2a^2}{r}+\frac{a^4}{r^3}\right)\sin(2\theta)\right]$$

$$(2)$$

At the hole wall, $r=a$ and

$$u = \left(\frac{1-v^2}{E}\right)\left[\left(\frac{S_x+S_y}{2}\right)(2a)+\left(\frac{S_x-S_y}{2}\right)(4a)\cos(2\theta)\right]$$
$$v = \left(\frac{1-v^2}{E}\right)\left[-\left(\frac{S_x-S_y}{2}\right)(4a)\sin(2\theta)\right]$$

$$(3)$$

where an outward radial displacement is positive and counter-clockwise circumferential displacement is also positive. Note: Tensile stresses are also positive and Equations (2) and (3) are for plane strain conditions. Equations (2) and (3) require modification if the hole is pressurized.

An interesting feature of solutions to plane problems in the mathematical theory of elasticity is the absence of elastic properties in the equations for stress. Thus, Equations (1) are valid for any homogeneous, isotropic, linearly elastic material. However, elastic properties do appear in the equations for displacement as seen in Equations (2) and (3).

A test for the validity of the stress and displacement solutions would be to differentiate the displacements to arrive at expressions for the associated strains, and then to substitute the strains into Hooke's law to obtain expressions for the stresses. Differentiation of the stresses and substitution into the equations of equilibrium should then satisfy the stress equations of equilibrium. The stresses should also satisfy the stress boundary conditions at the inner boundary (hole wall) and the far boundary at infinity. For completeness, one notes that the z-direction displacement is nil (plane strain); no variation in the z-direction occurs. For this reason, the z-direction strains are nil and the z-direction shear stresses are nil, but not the z-direction normal stress (which is?).

1. Generate the files necessary for analysis of a circular hole excavated in an initially stressed, homogeneous, isotropic elastic rock mass using a mesh generator. Specify 18 elements per ring and 20 rings for a total of 360 elements. The mesh is shown in Figure 2 below. Recall that a circle is a special ellipse (aspect ratio of 1) should your mesh generator have an ellipse option.

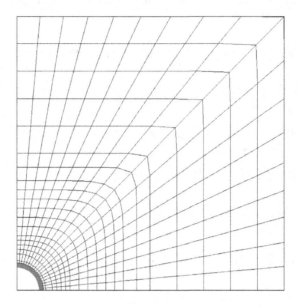

Figure 2 A plot of the circular hole mesh.

Notes:

(1) Use an initial stress state of $\sigma_{yy} = -1,000$ psi and all other stresses equal to zero. Use the sign convention that tension is positive.

(2) Use a Young's modulus of 2.4 (106) psi and a Poisson's ratio of 0.20.

(3) Use boundary conditions that prevent displacement perpendicular to the sides of the mesh at the mesh sides. Can mix, but force and displacement in the same direction are not allowed. Usually, displacement takes precedence and force is ignored when in conflict.

(4) Use either 4CST quadrilaterals which are explicitly integrated, or numerically integrated quadrilaterals.

(5) Proceed with excavation from an initial stress state as in construction practice.

2. Verify the computer program by comparing output with the known solution, that is,

(1) Plot radial and tangential normal stress concentrations as functions of angle from the horizontal *x*-axis using the radius to the center of the elements in the first ring of elements at the hole wall. Also plot the shear stress in the same manner.

Reply:

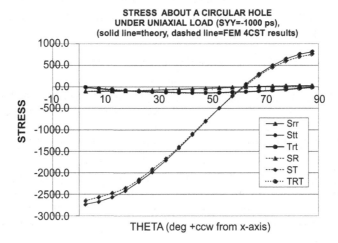

(2) Do a single regression analysis of all stress components from all elements in the first element ring on the known values. Plot the results and show the regression lines in an *x–y* format. Use the *x*-axis for known values and the *y*-axis for finite element output. Label the plot and indicate the regression parameters (correlation coefficient, slope, intercept, and the number of data points). Also place the regression equation on the plot (Figure 3).

   Reply:

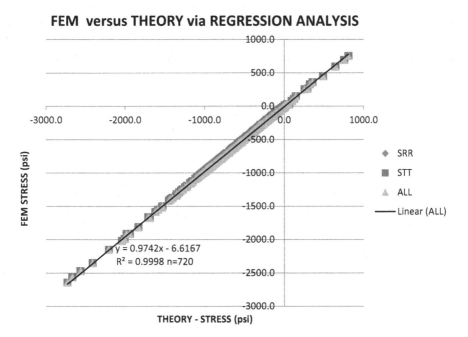

Figure 3 Regression of finite element radial and tangential stresses using the 4CST element on the same stresses from theory.

3. Compare 4CST results with N4Q (numerically integrated quadrilaterals) by plotting all the stresses from the 4CST output on the *x*-axis with N4Q results on the *y*-axis. Do a regression analysis and present the correlation coefficient and regression line equation on the plot. Discuss.

      Reply:

      The requested plot is shown in Figure 4.

      The origin of coordinates is at the lower left in the standard position. Quarter symmetry is implied with the boundary conditions imposed that prevent displacement normal to the sides of the mesh. Reference to the coordinates file shows the excavated hole has a radius of 3 to the interface between gray and white elements; width (and height) of the mesh is 30. The gray ring is a ring of small elements that are excavated (cut elements). In this regard, excavation changes cut elements to "air" elements of zero stress and then computes the out-of-equilibrium forces at the boundary of the air and rock elements. These forces are cut or excavation forces that are equal, but opposite in sense to the forces associated with preexcavation (initial) stresses; they load the adjacent rock and induce displacements, strains, and stress changes in the mesh.

GEOFEM 7 KEY

Axial symmetry arises in many applications, especially in analyses of stress about drill holes and well bores under gravity loading only. In such cases, the horizontal stresses before drilling the hole, say, relative to compass coordinates, x=east, y=north, z=up, are equal, that is, $S_{xx}=S_{yy}$. Relative to cylindrical coordinates, the prehole stress field is axially symmetric; that is, no variation with the polar angle occurs. Axial symmetry also prevails in hollow cylinders such as pipes under internal and external pressures.

In axial symmetric problems, the use of axially symmetric elements provides a substantial reduction in the number of elements needed to model the structure at hand,

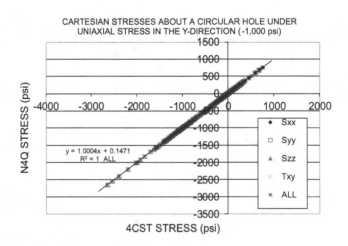

Figure 4 Regression of numerically integrated quadrilateral stresses about a circular hole with 4CST quadrilateral stresses.

for example, conventional elements approximately square formed at 5° increments about a circle would number 72. Only one element axially symmetric element would be needed.

Figure 1 shows a hollow cylinder under internal and external pressures.

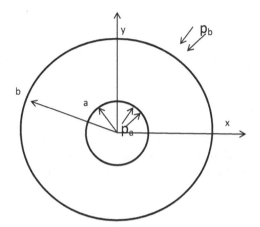

Figure 1 Diagram of a hollow cylinder under internal and external pressures.

In the elastic range of deformation, the radial and circumferential stresses are given by the formulas (*with compression positive*):

$$\sigma_r = p_b - (p_b - p_a)\left(\frac{1}{b^2} - \frac{1}{r^2}\right) \bigg/ \left(\frac{1}{b^2} - \frac{1}{a^2}\right) = \sigma_3$$

$$\sigma_\theta = p_b - (p_b - p_a)\left(\frac{1}{b^2} - \frac{1}{r^2}\right) \bigg/ \left(\frac{1}{b^2} - \frac{1}{a^2}\right) = \sigma_1 \qquad (1)$$

$$\sigma_z = v(\sigma_r + \sigma_\theta) = \sigma_2$$

where the ordering of principal stresses is based on the assumption that the internal pressure $p_a = 0$. These equations are based on equilibrium of stresses, strain–displacement relationships, and Hooke's law for a homogeneous, isotropic, linearly elastic hollow cylinder and the conditions of the problem: axial symmetry and plane strain.

Accordingly, stress equilibrium with neglect of weight reduces to

$$\partial\sigma_r / \partial r + (\sigma_r - \sigma_\theta)/r = 0 \qquad (2)$$

The strain–displacement equations reduce to

$$\varepsilon_r = \partial u / \partial r, \ \varepsilon_\theta = u/r \qquad (3)$$

Use of (3) in Hooke's law expresses stresses in terms of radial displacement u when substituted in (1), leading to an ordinary differential equation of second order that allows for direct integration and solution for u. The solution has the form:

$$u = C_1 r + C_2 / r \qquad (4)$$

Back-substitution of (4) into (3) and then into Hooke's law gives the stresses that after imposition of boundary conditions allows for determination of the two constants of integration and finally Formulas (1). The process is straightforward, but somewhat lengthy, where careful attention to algebraic signs is essential. Thus, with pressure positive,

$$\sigma_r = p_b\left(\frac{1-a^2/r^2}{1-a^2/b^2}\right) = \sigma_3$$

$$\sigma_\theta = p_b\left(\frac{1+a^2/r^2}{1-a^2/b^2}\right) = \sigma_1 \tag{5}$$

$$\sigma_z = v(\sigma_r + \sigma_\theta) = \sigma_2$$

The questions of where the first yield occurs and under what external pressure are answered by consideration of the yield condition or failure criterion, in this case, the well-known Drucker–Prager (DP) condition. This condition in terms of principal stresses is

$$\left\{(1/6)[(\sigma_1 - \sigma_2)^2 + (\sigma_2 - \sigma_3)^2 + (\sigma_3 - \sigma_1)^2]\right\}^{1/2} = A(\sigma_1 + \sigma_2 + \sigma_3) + B$$

that is, $\tag{6}$

$$J_2^{1/2} = AI_1 + B, \quad A = (1/\sqrt{3})(C_0 - T_o)/(C_o + T_o), \quad B = (2/\sqrt{3})(C_oT_o)/(C_o + T_o)$$

where C_o and T_o are unconfined compressive and tensile strengths.

Of interest is the sum of the radial and tangential stresses. Inspection of (5) shows this sum to be a constant, and because the intermediate principal stress is just Poisson's ratio times this same sum, the invariant I_1 is constant. Consequently, the right-hand side of (6) is a constant through the thickness of the cylinder. The left-hand side of (6) is a lengthy expression after substitution of (5), but interestingly, all terms vary with $1/r^2$. When the left-hand side is made equal to the right-hand side, the value of r is just the value necessary to cause first yielding. Values less than a are not admissible, of course. The first trial value of r is then just a. If r is increased beyond a, the left-hand side is diminished. Thus, first yielding must occur at $r = a$. Evaluation of (6) with $r = a$ then leads to the value of the externally applied pressure $p_b{}^*$ that causes first yielding. This value should be expressed in terms of material strengths and cylinder radii.

Further increase in p_b expands the yielding zone beyond a and ultimately to b when collapse must occur with vanishing of the elastic outer ring. The elastic–plastic solution should be found using the FEM with elastic–perfectly plastic capability and an option for the Drucker–Prager yield criterion that also serves as a plastic potential (associated flow rules).

1. Consider a hollow cylinder with an outer radius b twice the inner radius a. Further suppose the cylinder ends are fixed in the axial direction, so overall a plane strain condition prevails. Apply a uniform radial pressure p_b against the outer wall of the cylinder ($r = b$).

    Use the *elastic* solution to this problem and a Drucker–Prager yield condition to determine: (1) where the first yielding occurs and (2) under what applied external pressure. Use an unconfined compressive strength of 165.5 MPa (24,000 psi) and a tensile strength of 16.55 MPa (2.400 psi) for a strength ratio of 10.0 to determine the constants A and B. Then determine the pressure $p_b{}^*$ necessary to just cause yielding at the inner wall of the cylinder $r = a$. Use a Young's modulus of 16.55 GPa (2.4 million psi) and a Poisson's ratio of 0.2. Assume isotropy.

Reply:

$$A = \left(\frac{1}{\sqrt{3}}\right)\left(\frac{C_o/T_o - 1}{C_o/T_o + 1}\right), \qquad B = \left(\frac{2}{\sqrt{3}}\right)\left(\frac{C_o}{C_o/T_o + 1}\right),$$

$$= \left(\frac{1}{\sqrt{3}}\right)\left(\frac{10-1}{10+1}\right) \qquad\qquad = \left(\frac{2}{\sqrt{3}}\right)\left(\frac{24,000}{10+1}\right)$$

$$\underline{A = 0.472} \qquad\qquad\qquad \underline{B = 2,519 \text{ psi } (17.375 \text{ MPa})}$$

$$\sigma_r(r = a) = p_b - (p_b)\left(\frac{1}{b^2} - \frac{1}{a^2}\right)\Big/\left(\frac{1}{b^2} - \frac{1}{a^2}\right) = \sigma_3$$

$$\sigma_r(r = a) = 0 = \sigma_3$$

$$\sigma_\theta(r = a) = p_b - (p_b)\left(\frac{1}{b^2} + \frac{1}{a^2}\right)\Big/\left(\frac{1}{b^2} - \frac{1}{a^2}\right) = \sigma_1$$

$$\sigma_\theta(r = a) = p_b\left(\frac{2}{(a/b)^2 - 1}\right)$$

$$\underline{\sigma_\theta(r = a) = -2.67\, p_b = \sigma_1 \text{ (compression+)}}$$

$$\sigma_z(r = a) = v(\sigma_r + \sigma_\theta) = \sigma_2$$

$$\underline{\sigma_z(r = a) = 0.2(0 - 2.67\, p_b) = -0.533 = \sigma_2}$$

$$\left\{(1/6)[(\sigma_1 - \sigma_2)^2 + (\sigma_2 - \sigma_3)^2 + (\sigma_3 - \sigma_1)^2]\right\}^{1/2} = A(\sigma_1 + \sigma_2 + \sigma_3) + B$$

$$\left\{(1/6)\left[(-2.67\,p_b - (-0.533\,p_b))^2 + ((-0.533\,p_b) - 0)^2 + (0 - -2.67\,p_b)^2\right]\right\}^{1/2}$$

$$= 0.472(-2.67\,p_b - 0.533\,p_b + 0) + 2519$$

$$1.413\,p_b = 1.512\,p_b + 2519 \;\therefore$$

$$\underline{p_b = -25,444 \text{ psi } (-175.48 \text{ MPa}) \quad \text{(pressure is negative)}}$$

where the accuracy is subject to round off during the calculation. When done in a spreadsheet, the result is $p_b = 25,058$ psi (172.8 MPa).

2. Generate a mesh with a ratio of outer to inner radii of 2, and use 10 axially symmetric square elements for the finite element mesh. For example, $a = 10$ and $b = 20$. Displacement in the axial z-direction should be prevented to ensure plane strain conditions in consideration of a long hollow cylinder. The two node forces at $r = b$ are equivalent to a uniformly distributed pressure on the outer surface of the cylinder section. The total force is (pressure) (area), of course. But recall that elements are rings, so the area of pressure application A is $A = (2\pi b)(1)$. Note the element is one unit thick in the z-direction. Apply forces computed from the pressure just necessary to cause yielding at the inner wall of the cylinder.

Reply:

The mesh is shown in Figure 2.

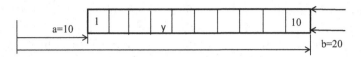

Figure 2 Ten axially symmetric elements, square in cross section with node forces applied in consideration of an external pressure at *b*.

Compare the FEM results with the analytical results in the elastic range to the point where the first yield appears. Make the comparison graphically in by plotting the radial, tangential, and axial stresses across the cylinder. Plot theoretical stress (radial, tangential, and axial) along the *x*-axis and FEM results along the *y*-axis. Then do a regression analysis of FEM results (all three stresses for a point total of 30) on theory through the cylinder.

Reply:

Figure 3 is the requested plot.

Figure 4 is the plot of FEM stresses versus theory and a regression of all FEM stresses on theory. The correlation is excellent as seen in the regression line equation and the correlation coefficient (*R*). The ten elements produce 30 values of stress.

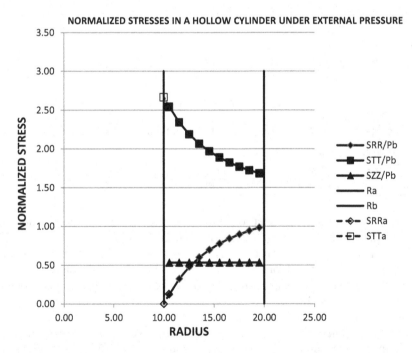

Figure 3 Stresses across a hollow cylinder under a uniform external pressure. Data are from FEM. Stresses are normalized through division by the externally applied pressure P_b and plotted at the centers of the ten elements through the cylinder. Extrapolation of stress from the element centers to the hole wall would produce an estimate of stress at the hole wall. This is done in Figure 3.

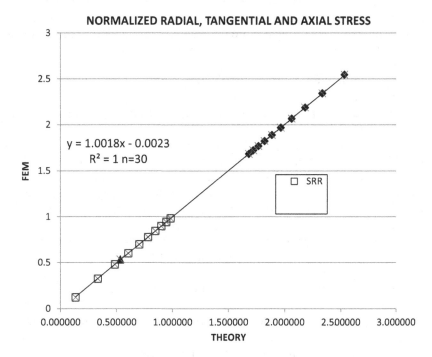

Figure 4 FEM stresses in a hollow cylinder in relation to stresses from theory and a regression of FEM on theory stresses (radial, tangential, and axial) in normalized form.

A plot of element safety factors is shown in Figure 5. The color scale indicates the safety factor. In the first element on the right that is near the hole wall, the factor of safety is 1.6 by the color scale and 1.66 by the finite element output that also produces element safety factors as well as element stresses. These values are average values of stresses in the element.

An element safety factor *fs* is defined as $fs = \left(J_2^{1/2}(\text{maximum}) / J_2^{1/2}(\text{actual}) \right)$. A safety factor greater than 1 indicates an element in the elastic domain. A safety factor of 1 indicates the limit to elasticity. Subsequently, loading would result in elastic–plastic deformation. A safety factor less than one is not allowed, but may occur during numerical analyses.

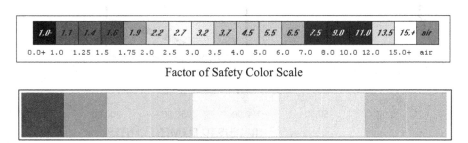

Factor of Safety Color Scale

Figure 5 Element safety factors across the 10-element hollow cylinder mesh.

Repeat the analysis with a mesh of 100 square elements across a cylinder with an inner radius of 100 and an outer radius of 200. The radius ratio (*b/a*) remains at two, but the mesh is much more refined and thus enables a more detailed stress distribution. Also repeat the plots and regressions analysis.

Reply:

Figure 6 shows normalized FEM stresses through the cylinder using 100 elements. The density of points is quite evident in the figure, which suggests a more accurate estimate of stress concentration near the inner wall of the cylinder where yielding occurs first.

Figure 7 shows a plot of FEM stress versus stress from theory and a regression of FEM stress on theoretical stress including the three normal stresses through the cylinder. The correlation is excellent as evident in the correlation coefficient *R*.

3. Monitoring displacements is one method of determining safety of the cylinder or well bore. Determine the relative change (percent of initial diameter) in inner diameter of the cylinder under the external pressure just necessary to cause the first yielding at the well bore wall. Also determine the relative change in outer diameter at the first yield. Do this for the 10-element and 100-element meshes.

Reply:

In the case of the 10-element mesh, the radial displacements of inner and outer cylinder walls are −0.2672 and −0.2839 in., respectively (−0.679 and −0.721 cm). The negative sign indicates inward displacement. In the case of the 100-element mesh, the displacements are the same! Changes in diameter are just twice the radial displacements. In case of the 10-element mesh analysis, the relative percentage change

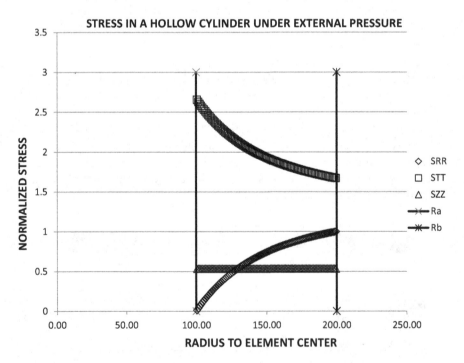

Figure 6 Normalized FEM stresses through the cylinder using 100 elements.

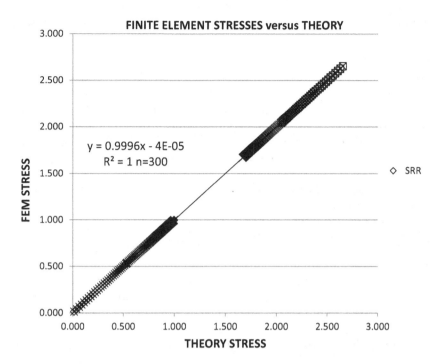

Figure 7 FEM stress versus theory and regression of FEM stress on theoretical stress.

in the inner diameter of the cylinder is 100 (0.534/20) or 2.672%. The relative change in the outer diameter (40 in.) is 1.420%. The percentage changes in the 100-element mesh case are just one-tenth those in the 10-element case because the diameters are ten times the 10-element cylinder diameters.

4. Discuss results and reach conclusions about axial symmetry and elastic–plastic analyses undertaken in this exercise. Be sure to address a question of mesh refinement and the question of accurate stress concentration at the limit to a purely elastic response, that is, at the first yield.

Reply:

Axial symmetry allows for an order of magnitude savings in elements required for analysis and therefore in computational effort. However, the meshes in this exercise were small and run times were in seconds, so the savings would be quite small relative to a full circle mesh.

The mesh refinement was productive in that a more accurate estimate of stress concentration is possible because the element center is closer to the inside wall in the case of the 100-element mesh. However, extrapolation of the plotted stress from the wall element center to the wall allows for an accurate estimate in geotechnical work such as well or drill hole analysis.

Estimates of diameter changes inside and outside a hollow cylinder would allow for judging the safety of the cylinder in consideration of the ease of measuring fractions of inches or centimeters. Such monitoring could also be used to estimate the applied pressure in case of a well or drill hole, the prehole stress field.

Indeed, measurement of borehole diameter changes is an important concept in stress measurement in situ, for example, with a borehole deformation gauge.

GEOFEM 8 KEY

This assignment explores nonlinear FEM stress analyses in two dimensions. The problem is to compute the stress, strain, and displacement fields about a long tunnel with a circular cross section excavated in a homogeneous, isotropic, linearly elastic material that yields in accordance with linear and quadratic forms of the well-known Drucker–Prager yield condition. The yield condition in case of isotropy is $J_2^{n/2} + AI_1 = B$, where

> J_2 = 2nd invariant of deviatoric stress,
> I_1 = 1st invarinat of stress,
> A, B = material constants, and
> n = an exponent from testing.

In this problem, $n = 1$ and 2.

Consider an initial stress state of $S_{xx} = 6.90\,\text{MPa}$ (−1,000 psi), $S_{yy} = 20.69\,\text{MPa}$ (−3,000 psi), $S_{zz} = 34.48\,\text{MPa}$ (−5,000 psi), $T_{yz} = 0$, $T_{zx} = 0$, and $T_{xy} = 0$, where x is horizontal, y is vertical, and z is along the tunnel axis (normal to the page).

This is a plane strain problem in view of the given conditions. A typical cross section and mesh are shown in the figure. This is the same mesh used in the previous elastic analyses.

Displacement normal to the sides is prevented; corners are fixed. Axes of symmetry in conjunction with sides far from the influence of the hole justify these boundary conditions.

Under plane strain conditions, the z-direction shear stresses and strains are nil. The z-direction normal stress may be computed through Hooke's law, i.e., $\sigma_{zz} = \nu(\sigma_{rr} + \sigma_{\theta\theta})$ in the elastic domain. However, once yielding occurs and the stress–strain relations become nonlinear, a numerical calculation is required.

Material properties for the material are as follows: Young's modulus = 17.93 GPa (2.6 million psi), Poisson's ratio = 0.26, unconfined compressive strength = 20.35 MPa (2,951 psi), tensile strength = 1.2 MPa (174 psi), and shear strength = 1.3 MPa (190 psi) in case $n = 1$. Shear strength in case $n = 2$ is 28.6 MPa (414 psi).

Use ten load increments with a maximum number of iterations of 2,000 each load step.

1. Generate the files necessary for analysis of a circular hole excavated in an initially stressed, homogeneous, isotropic elastic rock mass using a mesh generator. Specify 18 elements per ring and 20 rings for a total of 360 elements. The mesh is shown in Figure 2. Recall that a circle is a special ellipse (aspect ratio of 1) should your mesh generator have an ellipse option.

    Reply:
    The files necessary were generated as shown in Figure 1.
2. This exercise explores material nonlinearity in a FEM analysis using an elastic–perfectly plastic differential stress–strain law. The exploration begins with equation solving using the iterative Gauss–Seidel approach and continues with another

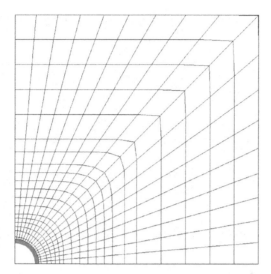

Figure I A plot of a circular hole mesh. Only one-quarter of the section is shown in consideration of symmetry with respect to the *x*- and *y*-axes.

popular iterative approach, a conjugate gradient solver. Comparisons based on the extent of the yielding zone about the tunnel, accuracy, and solution time are requested, with results compared in table format as in the elastic analyses problem. Table 1 is an example of stress comparison.

Table I GEOFEM 7 tunnel results

Approach/object	Gauss–Seidel	Conjugate gradient
Solution time (seconds)	8	4
Number of yielding elements	57	57
Stress in element near *x* S_{xx}, S_{yy}, S_{zz}, T_{yz}, T_{zx}, T_{xy}	−573, −7,201, −241, 0, 0, 262 psi −3.9, −49.7, −43.0, 0, 0, 1.8 MPa	−573, −7,201, −6241, 0, 0, 262 psi −3.9, −49.7, −43.0, 0, 0, 1.8 MPa
Stress in element near *y* S_{xx}, S_{yy}, S_{zz}, T_{yz}, T_{zx}, T_{xy}	294, −602, −1245, 0, 0, 11 psi 2.0, 4.1, −8.6, 0, 0, 0.1 MPa	294, 602, −1245, 0, 0, 11 psi 2.0, 4.1, −8.6, 0, 0, 0.1 MPa
Stress far from hole wall S_{xx}, S_{yy}, S_{zz}, T_{yz}, T_{zx}, T_{xy}	−42, −2,864, −4,976, 0, 0, 2 psi 0.3, 19.7, −34.3, 0, 0, 0 MPa	−42, −2,864, −4,976, 0, 0, 2 psi 0.3, 19.7, −34.3, 0, 0, 0 MPa

Reply:
   Results are presented in Table 1 as requested.
3. Show the extent of yielding elements in a plot such as in Figure 2.
       Reply:
       The extent of yielding elements is shown in Figures 2 and 3 as requested.

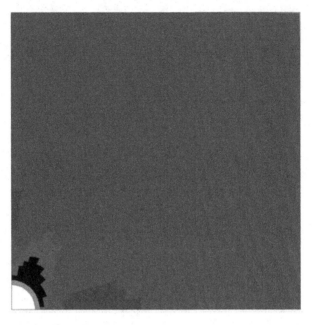

Figure 2 Extent of yielding (black) elements in case $n = 1$ (Gauss–Seidel and conjugate gradient analyses).

Figure 3 Extent of yielding (black) elements in case $n = 2$ (Gauss–Seidel and conjugate gradient analyses).

4. Discuss and reach conclusions from your comparison data in a short paragraph.
   Reply:
   Both iterative procedures produce identical results when the stresses, displacements, and distribution of yielding elements are examined. Thus, the important difference between the two methods is in the time required for solution. Solution time is mainly time spent in the equation solver. In this comparison, the conjugate gradient solver converges almost twice as fast as the Gauss–Seidel solver which used over-relaxation, to be sure.

   Yielding in the linear case ($n = 1$) is more extensive than in the quadratic case ($n = 2$) as seen in the figures. This result is somewhat surprising because the increase in strength with confining pressure is considered too high in the linear case according to many results from laboratory testing of rock cylinders under axial load and confining pressure. While this is true beyond some value of confining pressure, the reverse is true under low confinement as seen in these results. In both cases, failures are associated with compressive stress for the most part.

GEOFEM 9 KEY

This exercise concerns beam bending beyond the elastic range and invokes elastic–plastic finite element analysis considering Drucker–Prager yielding, but with the special provision of equal unconfined compressive and tensile strengths expected in case of mild steel beams used in structures such as bridges and buildings.

Use a mesh generation program to generate a beam as in the past assignment focused on beam action. Generate a beam consisting of 12×120 square elements. In this exercise, the beam should be simply supported and loaded with a point force at midspan. The analysis is plane strain and therefore two-dimensional with the x-axis along the beam and the y-axis vertical. In fact, the problem is one of sheet bending, but for simplicity, "beam bending" is used.

Use a Young's modulus of 16.55 GPa (2.4 (10^6) psi) and a Poisson's ratio of 0.20. Also use a Drucker–Prager yield condition in linear form ($n = 1$) with equal unconfined compressive and tensile strengths of 165.5 MPa (24,000 psi).

Use boundary conditions such that the node at the mid-height of the left-hand side of the beam (0, 6) is fixed in both directions (x and y) and the corresponding node at the right-hand side (120, 6) is fixed but only in the vertical y-direction. These boundary conditions correspond to a simply supported beam with end rotation possible.

1. Do a beam analysis and plot shear and moment diagrams.
   Reply:
   Note: These diagrams are determined from equilibrium requirements,
   $$\sum F = 0, \sum M = 0.$$

2. Determine the applied force necessary to cause first yielding and the location of the first yielding when plastic zones just appear using technical beam bending theory. Apply a force that is 90% of the load that would just cause yielding, so the deformation is purely elastic. Then compare the finite element distribution of stress at mid-span with the theoretical solution based on technical beam bending

theory. Note: This problem is slowly convergent, so when using an iterative solution, many iterations are necessary, perhaps 100,000.

Reply:

A simply supported beam loaded at mid-span by a point load is considered. Elastic and strength properties are given. Because tensile and compressive strengths are equal, the specified Drucker–Prager yield criterion ($n=1$) reduces to:

$$\left\{ (1/6)\left[(\sigma_1 - \sigma_2)^2 + (\sigma_2 - \sigma_3)^2 + (\sigma_3 - \sigma_1)^2 \right] \right\}^{1/2} = T/\sqrt{3} \tag{1}$$

In consideration of bending under plane strain conditions with z normal to the page, x along the beam, and y vertical, the principal stresses are $\sigma_1 = \sigma_x, \sigma_2 = \sigma_z = v\sigma_x, \sigma_3 = \sigma_y = 0$. Substitution into (1) leads to an estimate of the bending stress at first failure. Thus, from (1), $\left(1/\sqrt{3}\right)(\sigma_1)\left(1 - v + v^2\right)^{1/2} = \left(1/\sqrt{3}\right)T$, so $\sigma_1 = \sigma = 180.6$ MPa (26,186 psi) after substitution for Poisson's ratio and tensile strength.

The applied force necessary to cause first yielding may now be computed from beam theory and the stress at first yield. From beam theory and analysis of the problem beam

$$\sigma = Mc/I \text{ and } M = FL/4 \tag{2}$$

Hence, $F = 4\sigma I/Lc$. After substitutions, $F = 93.85$ kN (20,948 lbf), the force necessary to cause first yield.

Comparison of this estimate with finite element results requires further consideration. As before, the element stresses are averages at element centers, whereas beam theory gives us peak stresses at the beam surfaces. Thus, the force necessary to cause first yield in finite element analysis in this problem is the force necessary to cause yielding at $y = -5.5$ in. (the beam bottom at mid-span). The plastic zone should extend through the outer elements. The moment that occurs after yielding is given by the formula.

$$M = \left(\sigma h^2/4\right)\left[1 - (4/3)\left(\frac{c}{h}\right)^2 \right]$$

$$= \left(26,186 \,(12^2/4)\right)\left[1 - (4/3)(5/12)^2 \right] \tag{3}$$

$M = 678,741$ lbf-in.

The associated force in the problem is then $F = (678,741)\,(4)/120) = 22,624$ lbf (101.4 kN). This force in theory causes yielding to extend to $c = 5$ in. off the neutral axis. If the element centroid of 5.5 in. is used, then the force is a smaller estimate. The force in the finite element analysis to the bottom center element to yield will be somewhat in between these estimates that vary because of element thickness and stress averages at the element centers.

Figure 1 shows the distribution of element safety factors in the beam mesh under application of a point force at mid-span just sufficient to cause yielding at the

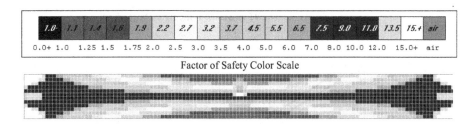

1.0-	1.1	1.4	1.6	1.9	2.2	2.7	3.2	3.7	4.5	5.5	6.5	7.5	9.0	11.0	13.5	15.+	air
0.0+	1.0	1.25	1.5	1.75	2.0	2.5	3.0	3.5	4.0	5.0	6.0	7.0	8.0	10.0	12.0	15.0+	air

Factor of Safety Color Scale

Figure 1 Element safety factor distribution according to the color bar under a point load at mid-span just sufficient to cause failure at the beam top and bottom surfaces.

beam top and bottom surfaces. The lowest element safety factor is 1.12 at the beam bottom, mid-span, as expected. At 90%, the applied force is 91.22 kN (20,361 lbf).

Figure 2 shows element safety factors when the applied load is 5% greater than the load sufficient to cause yielding to $c = 5$, that is, through the first layers of elements at the top and bottom of the beam. This result is close to technical beam theory. However, technical beam theory is an approximation that is only accurate to a few percent. In consideration of technical beam theory accuracy, the FEM result is actually quite good. In this regard, consider the stress distribution in the near vicinity of the load point and the support points. FEM gives accurate results in this regions, whereas technical beam theory does not.

Figure 3 is a graphical comparison of stress from FEM with theory.

Figure 4 is a regression analysis of FEM stress on theory. Excellent agreement is observed with a high correlation coefficient R.

3. Increase the applied force and rerun the analysis several times until sufficient data are available to plot a force–displacement curve that shows the elastic range and subsequent elastic–plastic behavior. Do the analysis using 4CST and N4Q (numerically integrated) linear displacement, constant-strain elements.

Reply:

While making a plot of beam force versus displacement, a puzzle arose because the plot was steepening as more elements failed when a decreasing slope was expected. The plot is shown in Figure 5. A steepening curve beyond the elastic limiting is a special material behavior known as "locking" behavior. In reviewing previous elastic–plastic beam analyses, the plot was as expected with a decreasing slope and rapidly increasing displacement as the applied force was increased beyond the elastic limit. Convergence slowed as the number of yielding elements increased, as one might expect. The locking behavior (diamond symbol) was obtained using numerically integrated quadrilaterals (4NQ elements). When the

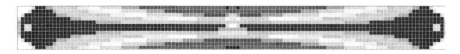

Figure 2 Element safety factor distribution according to the color bar under a point load at mid-span sufficient to cause failure through the first layers of elements at the beam top and bottom.

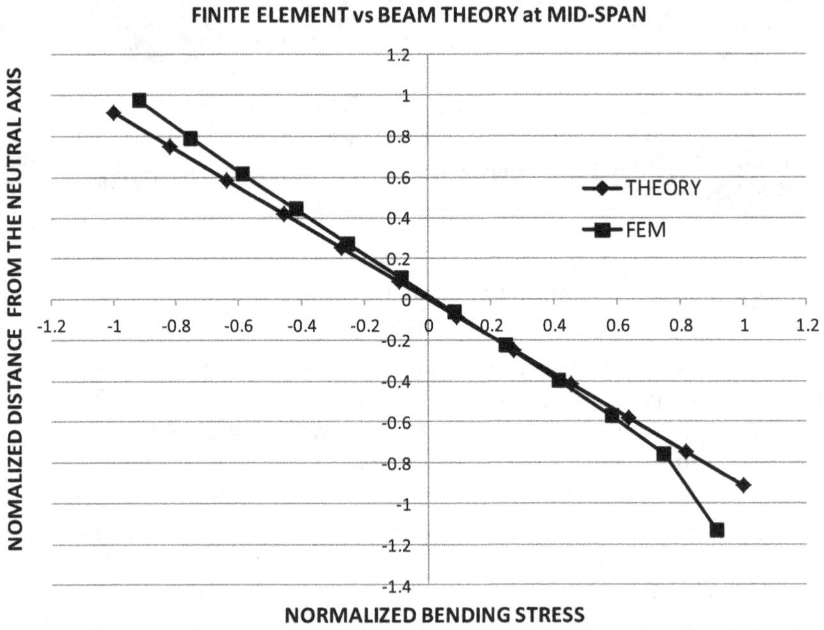

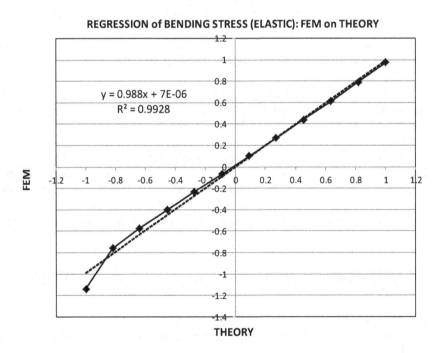

FINITE ELEMENT vs BEAM THEORY at MID-SPAN

Figure 3 Bending stress through the beam at mid-span. Bending stress is normalized through division by material strength. Distance from the neutral axis is normalized with division by the beam thickness. Positive values are above the neutral axis.

REGRESSION of BENDING STRESS (ELASTIC): FEM on THEORY

$y = 0.988x + 7E\text{-}06$
$R^2 = 0.9928$

Figure 4 Regression analysis of finite element results on theory. Data are from Figure 3.

program returned to quadrilaterals composed of four constant-strain triangles (4CST element), the plot (square symbols) in Figure 5 was obtained. In this regard, the 4CST element is a well-behaved element in plasticity analyses; not all elements are. An important and much referenced paper on the subject is the one by Nagtegaal, Parks, and Rice (1974).

Results for obtaining the collapse load from the finite element analysis were disappointing, as inspection of Figure 5 shows. The theoretical line should be horizontal. Although departure from linearity is evident at lower loads, the rise in the plot needs explanation. One explanation is in the fewer elements available for representing the elastic core of the beam as the plasticity regions spread toward the neutral axis. In essence, the mesh becomes much coarser as loading continues. When the plastic zone is at normalized $c = 1.0/12.0$ (above and below the neutral axis), the elastic core is represented by just two elements and is insufficient to capture the linear distribution of stress in the elastic core. Bending involves a second derivative of displacement, so second-order quadratic elements are preferred for beam analysis. Linear displacement, constant-strain elements are useful, but many more are needed for beam analysis. Our study of the cantilever beam indicated the same conclusion that stiff elements (triangles), or equivalently, too few elements, make the mesh too stiff for accurate beam analyses.

When the plastic zone extends through the beam to the neutral axis from top and bottom surfaces, collapse is impending because of the vanishing of the elastic core ($c = 0$). The associated force (normalized) from (3) is $F = 1.5$. In this regard, once yielding occurs, a purely elastic solution will no longer suffice. An

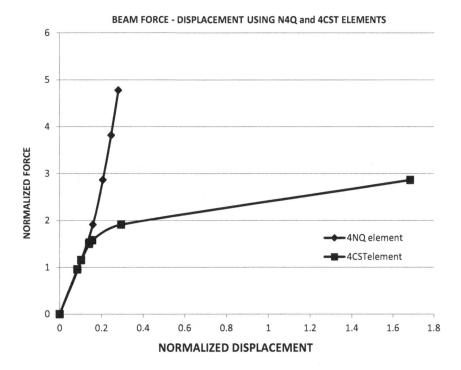

Figure 5 Force–displacement plots for two types of elements.

incremental loading using perhaps ten steps would be appropriate. The parameters in the run stream should be set accordingly. I used ten steps in my analyses and, of course, $n = 1$. As in the previous beam exercise, when the iterative equation solver is specified, a large number of iterations are necessary to reduce residuals by three orders of magnitude, a useful guide. In fact, 100,000 iterations were possible in each of ten steps.

4. Determine the applied force necessary to cause collapse when the plastic zones extend through the beam section.

   Reply:

   The applied force necessary to cause beam collapse in the finite element analysis was several multiples of the theoretical load. In my opinion, the reason is the approximation in technical beam theory is not what it should be. The departure of finite element analysis from theory is especially evident in the vicinity of the applied force. Beam theory takes no account of the complicated distribution of stress below a point load, although finite element analyses do. These observations were pointed out in class and would naturally enter into the discussion of results in the homework report.

REFERENCES

Hibbleler, R.C. (2008) *Mechanics of Materials* (7th ed.)., Upper Saddle River, NJ, Pearson Prentice-Hall, p 364.

Nagtegaal, J.C., D.M. Parks and J.R. Rice (1974) On Numerically Accurate Finite Element Solutions in the Fully Plastic Range. *Computer Methods in Applied Mechanics and Engineering*. Vol. 4, pp 153–177.

GEOFEM 10 KEY

A useful feature of FEM is the capability of using elements of different types in the same model. For example, elements representing beams can be used with plate elements in structural models. This exercise considers a simple bar element representing a rock bolt for rock reinforcement in models of underground excavations. This element is also known as a truss bar element.

Consider a bar element in one dimension illustrated in Figure 1. The element stiffness matrix $[k]$ may be obtained by first obtaining the element force–displacement relationship along the bar. The bar is assumed to be elastic with area A, length L, and Young's modulus E, so that

$$\sigma = E\varepsilon, \; f = \sigma A, \; d = L\varepsilon. \text{ Hence } f = (AE/L)d, \text{ so } k = AE/L, \tag{1}$$

Figure 1 Basic bar element.

where f is the force and d is the relative displacement between ends, that is, $f = kd$ much like a linear spring.

From (1),

$$f1 = k(d1 - d2) \text{ and } f2 = k(d2 - d1) \tag{2}$$

where $(d1 - d2)$ is the displacement of the point 1 relative to 2, and $(d2 - d1)$ is the displacement of point 2 relative to 1. In matrix form, (2) is

$$\begin{Bmatrix} f1 \\ f2 \end{Bmatrix} = \begin{bmatrix} k & -k \\ -k & k \end{bmatrix} \begin{Bmatrix} d1 \\ d2 \end{Bmatrix} \text{ or } \{f\} = [k]\{d\} \tag{3}$$

Consider the same bar element that responds only to axial load in three space dimensions (xyz) as illustrated in Figure 2. The element is represented by the heavy arrow and may have displacements u, v, and w in the x-, y-, and z-directions. Because the element responds only to axial load, the resultant displacement of the element is either elongation or shortening of the truss bar (no bending).

If C_x, C_y, and C_z are the direction cosines of the element, then the displacements of the element ends $d1x$, $d1y$, and $d1z$ and those similarly for point 2 in the x-, y-, and z-directions are

$$d1x = C_x d1, \ d1y = C_y d1, \ d1z = C_z d1 \text{ and } d2x = C_x d2, \ d2y = C_y d2, \ d2z = C_z d2 \tag{4}$$

Similarly for the forces

$$f1x = C_x f1, \ f1y = C_y f1, \ f1z = C_z f1 \text{ and } f2x = C_x f2, \ f2y = C_y f2, \ f2z = C_z f2 \tag{5}$$

In matrix form, (4) and (5) are

$$\begin{Bmatrix} d1x \\ d1y \\ d1z \\ d2x \\ d2y \\ d2z \end{Bmatrix} = \begin{bmatrix} C_x & 0 \\ C_y & 0 \\ C_z & 0 \\ 0 & C_x \\ 0 & C_y \\ 0 & C_z \end{bmatrix} \begin{Bmatrix} d1 \\ d2 \end{Bmatrix} = [C]\{d\} \text{ and } \begin{Bmatrix} f1x \\ f1y \\ f1z \\ f2x \\ f2y \\ f2z \end{Bmatrix} = \begin{bmatrix} C_x & 0 \\ C_y & 0 \\ C_z & 0 \\ 0 & C_x \\ 0 & C_y \\ 0 & C_z \end{bmatrix} \begin{Bmatrix} f1 \\ f2 \end{Bmatrix} = [C]\{f\} \tag{6}$$

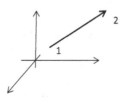

Figure 2 Bar in three dimensions.

Interestingly, the products $[C]^t[C] = [C][C]^t = [I]$, a 2×2 identity matrix. Thus,

$$\{f\}_{6\times1} = [C]_{6\times2}\{f\}_{2\times1} = [C]_{6\times2}[k]_{2\times2}[C]^t_{2\times6}\{d\}_{6\times1}, \text{ that is, } \{f\}_{6\times1} = [k]_{6\times6}\{d\}_{6\times1} \quad (7)$$

The last equation is the truss bar element stiffness equation in three-dimensional space. This transformed element stiffness matrix is a 6×6 array where each of the two nodes have three displacements. In component form, (7) is

$$\begin{Bmatrix} f1x \\ f1y \\ f1z \\ f2x \\ f2y \\ f2z \end{Bmatrix} = (k) \begin{bmatrix} C_xC_x & C_xC_y & C_xC_z & -C_xC_x & -C_xC_y & -C_xC_z \\ C_yC_x & C_yC_y & C_yC_z & -C_yC_x & -C_yC_y & -C_yC_z \\ C_zC_x & C_zC_y & C_zC_z & -C_zC_x & -C_zC_y & -C_zC_z \\ -C_xC_x & -C_xC_y & -C_xC_z & C_xC_x & C_xC_y & C_xC_z \\ -C_yC_x & -C_yC_y & -C_yC_z & C_yC_x & C_yC_y & C_yC_z \\ -C_zC_x & -C_zC_y & -C_zC_z & C_zC_x & C_zC_y & C_zC_z \end{bmatrix} \begin{bmatrix} u1 \\ v1 \\ w1 \\ u2 \\ v2 \\ w2 \end{bmatrix}$$

$$(8)$$

Consider a truss bar element parallel to the z-axis such that the direction angles from x, y, and z are $90°$, $90°$, and $0°$, respectively. The direction cosines C_x, C_y, and C_z are then 0, 0, and 1, respectively. The only non-zero matrix elements in (8) are the C_zC_z elements. Hence,

$$f1z = k(w1 - w2) \text{ and } f2z = k(-w1 + w2) \quad (9)$$

that has the requisite form (2). If the element were aligned with the x-axis, then the direction angles would be $0°$, $90°$, and $90°$, respectively, and C_x, C_y, and C_z would be 1, 0, and 0. According to (8) then with only C_xC_x not zero

$$f1x = k(u1 - u2) \text{ and } f2x = k(-u1 + u2) \quad (10)$$

Once forces and displacements are determined, stress and strains follow from (1).

A simplification is possible in practice by equating k to an "effective" Young's modulus E_b and thus eliminating any consideration of A and L separately. If the element were a real bar, then an effective modulus would not be appropriate. Otherwise, the element can be used to prevent overlap by adjacent surfaces. This usage occurs in simple contact algorithms that allow surfaces to move close together using a very low effective modulus value until contact is imminent when a very high modulus value is brought into play. The corresponding stress–strain plot for the truss bar would have the appearance in Figure 3. The abrupt transition in Figure 3 could be smoothed to the dotted line, which is typical of uniaxial tests on sand fill used in some mines. Such a curve may also be used to mimic compaction of caved ground formed in the course of some longwall mines. However, the curve is one-dimensional and use in finite element modeling requires a proper stress–strain relationship that links all stresses and strains for the simple reason that fill and caved material are not compacted under uniaxial load in actual practice. Indeed, the question of fill and caved

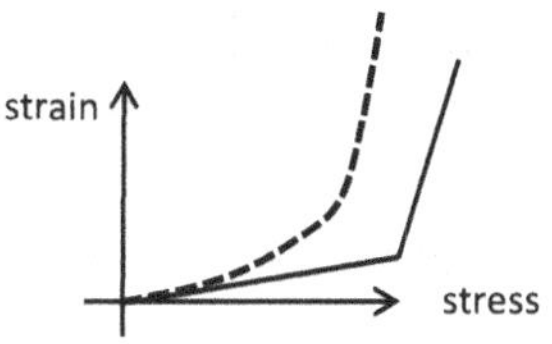

Figure 3 Stress–strain plot for overlap prevention.

material modeling goes well beyond consideration of a simple one-dimensional truss bar element.

Truss bar elements find use in assembled trusses illustrated in a simple manner in Figure 4, where all elements are pin-connected and thus do not support moments at the connections. The truss bar elements are thus only loaded axially. Each member of the truss can be represented by several truss bar elements, and the truss can be much more complicated in three dimensions. The truss members could be steel of various shapes, wide flange, box, and so forth. When the members are not considered pin-connected, then the elements should be beam elements that take into account bending moments as well as axial loads.

1. Consider a steel bar 16 mm (5/8 in.) in diameter and 1.83 m (6 ft) in length with a Young's modulus of 200 GPa (29 million psi) and a Poisson's ratio of 0.2. The tensile strength is 517 MPa (75,000 psi), and the ultimate strength is 688 MPa (100,000 psi). These properties correspond to a conventional rock bolt used to support rock in the vicinity of underground excavations. Note: The ductility of steel suggests equal compressive and tensile strengths.

Compute the bolt (bar) stiffness k_b. Also compute the axial stiffness k_a of a 4CST element 1.83 m (6 ft) square and the axial stiffness k_c for a N4Q element of the same dimensions. Plot force–displacement curves that show axial response in the elastic range of all three elements. Hint: Repeated single-element runs of the 4CST and N4Q elements will be necessary to gather the necessary force–displacement plots. Five points (displacement, force) for each plot in the elastic domain are suggested. These data should be tabulated before plotting.

Reply:
$$k = AE/L$$
By definition, $= \left[(\pi)(5/8)^2/4\right](29)(10^6)/(6)(12)$
$$k = 1.24\,(10^5)\ \text{lbf/inch}\ [21.8\ \text{kN/mm}]$$

Data from finite element testing are given in Table 1. In this regard, the data are identical for both types of elements (explicitly 4CST and numerically integrated N4Q) in this two-dimensional, plane strain experiment.

Figure 4 A simple truss composed of truss bar elements.

Table I Data from FEM testing for stiffnesses
of 4CST and N4Q elements

Force (N)	Displacement (mm)
322,560	0.73152
645,120	1.46304
967,680	2.19456
1,290,240	2.92608
1,612,800	3.6576

4CST and N4Q data are the same.

Figure 5 is a plot of data from Table 1. The slope of the regression line is the stiffness associated with the elements. Stiffness is 441 kN/mm [2.5 (10^6 lbf/in.)]. Thus, element stiffness is roughly twice the truss bar stiffness.

In engineering practice, rock bolts may be spaced on 6 ft (1.83 m) centers as a practical matter. If the test element were a three-dimensional element 6×6×6 ft (1.83×1.83 ×1.83 m), then the element stiffness would be 72 times the two-dimensional element stiffness, that is, 31.75 (10^6) N/mm [180(10^6) lbf/in.]. A three-dimensional analysis using a numerically integrated element verifies this inference. The table presents the data, and Figure 6 is a plot of the data. The result is a stiffness of 32.98 (10^6) N/mm [187(10^6) lbf/in.]. The agreement between two- and three-dimensional results is roughly 4%.

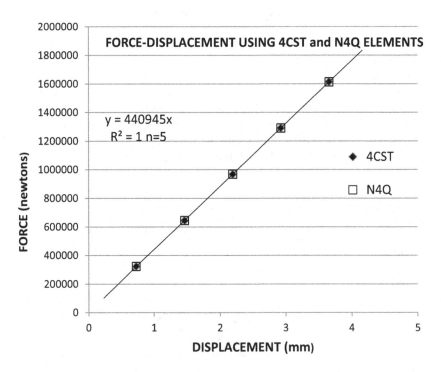

Figure 5 Force–displacement plot for 4CST and N4Q elements.

Table 2 Data from FEM testing in three dimensions

Force (N)	Displacement (mm)
0	0
2.3224E+07	0.705866
4.6449E+07	1.40716
6.9673E+07	2.11328
9.2897E+07	2.80924
1.1612E+08	3.52044

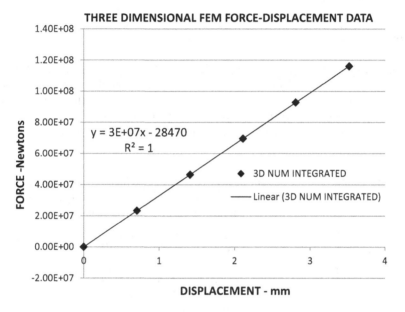

Figure 6 Force–displacement plot for a numerically integrated brick-shaped element. Data are from Table 2.

Rock bolt stiffness compared with element stiffness is shown to be orders of magnitude smaller than element stiffness and thus can do very little to inhibit rock wall displacement in actual practice, unless the "rock" has a very small Young's modulus, a modulus more soil-like than even jointed rock. This result indicates that rock bolts function in a manner that differs from direct inhibition of displacement. Conventional wisdom, indeed, states bolts prevent rock block falls and function by supporting rock blocks that may become detached from the parent rock mass. Thus, rock bolt design is properly based on support of dead weight loads. The rock bolts considered here would support a dead weight load of 103 kN (23,000 lbf), which is roughly equivalent to a rock block 1.8×1.8 m square and 1.2 m thick (6×6 ft square and 4 ft thick). So pattern rock bolting in underground rock excavations is an effective rock reinforcement method. Indeed, rock bolts are widely used in industry.

An issue of element assembly occurs when mixing elements of different types. However, the principle remains the same: The total node force is the sum of forces contributed to the node by elements that share the given node. Thus,

$$F_i = \sum_e f_i = \sum_e \sum_{j=1}^{j=n} k_{ij}\delta_j.$$

where e indicates elements that share node i and n ranges over the number of nodes an element has, two in case of a bar element and three or four in case of triangles or quadrilaterals.

GEOFEM II KEY

1. Use the divergence theorem in the form of "a principle of virtual power", to derive a continuum expression for slow, incompressible fluid flow in a rigid, porous medium governed by Darcy's law. Use P for fluid stress with compression negative and v for fluid displacement relative to the porous solid. Use $\dot{w}$ for fluid velocity, and note the displacement and velocity are vectors. Do the derivation in xyz coordinates.

    Reply:

    The divergence theorem in the form of a principle of virtual power equates the rate of work done by the external forces to the rate of change of internal energy. Beginning with the power of surface pressure, one has via the divergence theorem

$$\int_A Pv_n\,dA = \int_V \left[\partial(Pv_x)/\partial x + \partial(Pv_y)/\partial y + \partial(Pv_z)/\partial z\right]dV$$
$$= \int_V \left[P\partial(v_x)/\partial x + P\partial(v_y)/\partial y + P\partial(v_z)/\partial z\right]dV$$
$$+ \int_V \left[v_x\,\partial P/\partial x + v_y\,\partial P/\partial y + v_z\,\partial P/\partial z\right]dV$$
$$\int_A Pv_n\,dA = \int_V \left[v_x\,\partial P/\partial x + v_y\,\partial P/\partial y + v_z\,\partial P/\partial z\right]dV \qquad (1)$$

The first term in (1) is the work rate of pressure forces on the velocity normal to the surface A. The fluid velocity normal to the bounding surface A is the inner product $\{v\} \bullet \{n\} = v_x n_x + v_y n_y + v_z n_z$, where $\{n\}$ is an outward unit normal vector to A. The first integral in the second line of (1) is zero in consideration of incompressibility implicit in Darcy's law. Note: The sum of the velocity derivatives is the rate of dilatation or volume change and thus zero.

In matrix notation, (1) is

$$\int_A (P)\{v\}^t\{n\}\,dA = \int_V \begin{Bmatrix} \partial P/\partial x \\ \partial P/\partial y \\ \partial P/\partial z \end{Bmatrix}^t \{v\}dV \qquad (2)$$

where $\{\}$ indicates a 3×1 vector and $()$ indicates a scalar.

The work of the internal forces, gravity, is given by the integral

$$\int_V \{\gamma\}^t \{v\} dV \tag{3}$$

where γ is the specific weight of the fluid flowing. This integral appears on both sides of (1) and so can be omitted in the derivation.

2. Assume a linear interpolation of pressures in a finite element, and then using conventional notation, recast the continuum result in Problem 1 into finite element form. Be sure to define all terms and matrix dimensions.

   Reply:

   Linear interpolation over a triangle has the basic form of pressure P: $P = a_0 + a_1 x + a_2 y$, and thus,

$$P = N_1 P_1 + N_2 P_2 + N_3 P_3 \quad \text{In matrix form} \tag{4}$$

$$(P) = [N]\{P_n\} \tag{5}$$

where the Ns are the interpolation functions previously derived for deformation of solids and the Ps on the right-hand side are pressures at the nodes.

Pressure derivatives are obtained by differentiation of (4). Thus,

$$\begin{Bmatrix} \partial P/\partial x \\ \partial P/\partial y \\ \partial P/\partial z \end{Bmatrix} = [B_p] \begin{Bmatrix} P_1 \\ P_2 \\ P_3 \end{Bmatrix} \tag{6}$$

The matrix $[B_p]$ is analogous to the B-matrix that relates node displacements to strains (displacement derivatives) in case of solids.

Velocities are also interpolated such that

$$\{v\} = [N]\{v_m\} \tag{7}$$

where $\{v_m\}$ are the velocities at the three nodes of the considered triangle and $[N]$ is a matrix of interpolation functions as before.

The virtual power formulation is now

$$\int_A \{P_n\}^t [N]^t (q) dA = \int_V \{P_n\}^t [B_p]^t \{v\} dV \tag{8}$$

where (q) is the flow of fluid across the surface A per unit of area A, also known as the *specific discharge*.

The right-hand side of (8) can be refined further in consideration of Darcy's law. Thus,

$$\int_A \{P_n\}^t [N]^t (q) dA = \int_V \{P_n\}^t [B_p]^t [k] \left(\begin{Bmatrix} \partial P/\partial x \\ \partial P/\partial y \\ \partial P/\partial z \end{Bmatrix} + \{\gamma\} \right) dV \tag{9}$$

After rearrangement of terms, one has

$$\int_A [N]^t(q)dA - \int_V [B]^t[k]\{\gamma\}dV = \left(\int_V [B_p]^t[k][B_p]dV\right)\{P_n\}$$ (10)

in consideration that (9) must hold for arbitrary $\{P_n\}$ and constant terms in the $[B_p]$ matrix. In compact form, (10) is

$$\{f\} = [k']\{\delta\}$$ (11)

3. Verify the "Desai" example problem using a finite element seepage program (Desai and Abel). Also compute the true flow velocity, given a porosity n of 0.25. Note: By definition, $n = V_v/V$ and $V = V_v + V_s$ where subscripts v and s indicate void and solid, respectively.
   Reply:
   A simple finite element code SEEP written some years ago was used to solve this test problem with the specific weight of fluid given as 1 gm/cm^3. The program is three-dimensional and uses brick-shaped elements. There are four elements in this analysis and 18 nodes.
   The input run stream to SEEP is as follows:

```
      Seepage program trial Desai (brick 3D paris)
              Seepage program trial Desai (brick 3D paris)
      NUMBER OF ELEMENTS=   4
      NUMBER OF NODES=  18
      NODES WITH SPECIFIED HEAD OR VELOCITY=  12
      NUMBER OF MATERIAL TYPES=  1
      INTER= 10    MAXIT= 100    ERROR=0.0000E+00
      XFAC= 1.0    YFAC= 1.0    ZFAC= 1.0
            CONDUCTIVITY  PPOPERTIES
      NMAT =   1
          0.1000E+01      0.0000E+00      0.0000E+00
          0.0000E+00      0.1000E+01      0.0000E+00
          0.0000E+00      0.0000E+00      0.1000E+01
      ANGLX   ANGLY   ANGLZ
          0.0    0.0    0.0
      ELIMINATION SELECTED
          This is the end!
```

Nodes on the left- and right-hand sides have specified heads of 4 and 3 cm, respectively.
   Velocity results given at element centers are the following:

Elem	Mat	Xc	Yc	Zc	Vx	Vy	Vz
1	1	1.00	0.50	0.50	0.250	0.000	0.000
2	1	1.00	1.50	0.50	0.250	0.000	0.000

| 3 | 1 | 3.00 | 0.50 | 0.50 | 0.250 | 0.000 | 0.000 |
| 4 | 1 | 3.00 | 1.50 | 0.50 | 0.250 | 0.000 | 0.000 |

These results show a constant x-direction velocity of 0.25 cm/s through the four elements in agreement with results in "Desai". This is the Darcy or nominal velocity. The true velocity is faster. Given the volume flow rate $q = vA = v_t A_v$ computed using the nominal velocity and entire cross section of flow and by using the true velocity v_t and true area of flow (void area) A_t, the true velocity is $v_t = vA/A_v = v/n_A$, where n_A is the area porosity. This porosity is often assumed to be the conventional volume porosity n. Hence, in this case, the true velocity is $\begin{array}{c} v_t = 0.25/0.25 \\ \hline v_t = 1\ cm/s \end{array}$. The fluid moves four times faster than the nominal velocity.

Pressure heads at the 18 nodes in the mesh are as follows:

Node	X	Y	Z	UX
1	0.00	0.00	0.00	4.0000
2	0.00	1.00	0.00	4.0000
3	0.00	2.00	0.00	4.0000
4	2.00	0.00	0.00	3.5000
5	2.00	1.00	0.00	3.5000
6	2.00	2.00	0.00	3.5000
7	4.00	0.00	0.00	3.0000
8	4.00	1.00	0.00	3.0000
9	4.00	2.00	0.00	3.0000
10	0.00	0.00	1.00	4.0000
11	0.00	1.00	1.00	4.0000
12	0.00	2.00	1.00	4.0000
13	2.00	0.00	1.00	3.5000
14	2.00	1.00	1.00	3.5000
15	2.00	2.00	1.00	3.5000
16	4.00	0.00	1.00	3.0000
17	4.00	1.00	1.00	3.0000
18	4.00	2.00	1.00	3.0000

The interesting nodes are 4, 5, 6, 13, 14, and 15 with heads in the UX column. These nodes are in a vertical plane through the middle of the mesh as shown in Figure 1 by numbers 1, 2, and 3. Heads at these nodes are equal and have the value of 3.5 cm in agreement with "Desai".

The FEM results are in excellent agreement with "Desai" and thus give credibility to the program SEEP.

4. Extend the FEM analysis of seepage to accommodate compressibility of a fluid flowing in a compressible solid. Recall that the Darcy velocity is a velocity of the fluid relative to the solid velocity. Assume steady, saturated flow.

   Reply:

   Let the displacement of fluid and solid be U and u, respectively. Velocity of fluid relative to solid is then $\{v\} = \{\dot{U}\} - \{\dot{u}\}$. The rate of volume change of fluid relative to solid volume change is then

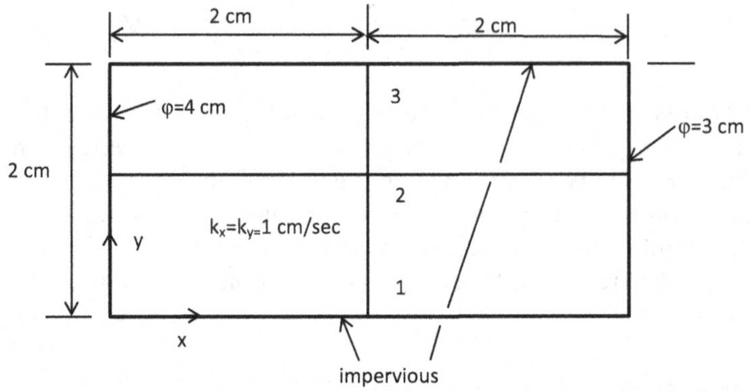

Figure 1 Mesh of four elements for a seepage analysis involving Darcy's flow. Hydraulic conductivities are given in the figure as are piezometric heads on the left- and right-hand sides of the figure. Top and bottom surfaces are impervious; no flow occurs across these boundaries. Note: φ is the piezometric head.

$$\begin{aligned} &\partial v_x/\partial x + \partial v_y/\partial y + \partial v_z/\partial z = \partial(\dot{U}-\dot{u})/\partial x + \partial(\dot{U}-\dot{u})/\partial y + \partial(\dot{U}-\dot{u})/\partial z \\ &\partial v_x/\partial x + \partial v_y/\partial y + \partial v_z/\partial z = \dot{\varepsilon}_v^f - \dot{\varepsilon}_v^s \end{aligned} \tag{12}$$

Hence, the first integral in the second line of (1) is no longer zero, and rather is given by (12), where the subscript v indicates volumetric strain and the superscripts f and s imply fluid and solid, respectively. The volumetric strain is related to stress via compressibility. Thus, $\dot{\varepsilon}_v^f = c\dot{P}$. And with a change in notation, $\dot{\varepsilon}_v^f - \dot{\varepsilon}_v^s = \partial\varsigma/\partial t$, where ς (zeta) is the change in fluid volume per unit total volume; the integral in question becomes

$$\begin{aligned} \int_V \left[P\partial(v_x)/\partial x + P\partial(v_y)/\partial y + P\partial(v_z)/\partial z) \right] dV &= \int_V (P)(\dot{\varsigma})dV \\ &= \int_V \left[(P)(c\dot{P} - \{c\}^t\{\dot{\varepsilon}\} \right] dV \end{aligned} \tag{13}$$

where $\{c\}^t = \{1\,1\,1\,0\,0\,0\}$ and $\{\dot{\varepsilon}\} = (\dot{\varepsilon}_{xx}^s\ \dot{\varepsilon}_{yy}^s\ \dot{\varepsilon}_{zz}^s\ 0\ 0\ 0)$; the product is simply the solid volumetric strain rate.

Frequently, a proportionality is assumed between fluid and volumetric strain rates, that is, $\dot{\varepsilon}_v^f \propto \dot{\varepsilon}_v^s$. This assumption allows for replacement of $\dot{\varepsilon}_v^s$ with constant times $\dot{\varepsilon}_v^f$ in (13) with the result

$$\int_V (P)(\dot{\varsigma})dV = \int_V (P)(S)(\dot{P})dV \tag{14}$$

where S is a "storage" coefficient. In FEM form, $\int_V (P)(S)(\dot{P})dV = \{P_n\}\left(\int_V [N]^t[S][N]dV\right)\{\dot{P}_n\}$.

The contribution to (10) is $\left(\int_V [N]^t [S][N] dV \right)\{\dot{P}_n\} = [k'']\{\dot{P}_n\}$. In shortened notation,

$$\{f\} = [k']\{\delta\} + [k'']\{\dot{\delta}\} \tag{15}$$

REFERENCE

Desai, C.S. and J.F. Abel. (1972). *Introductions to the Finite Element Method: A Numerical Method for Engineering Analysis*. Van Nostrand Reinhold Company, New York, pp 384–389.

GEOFEM 12 KEY

1. Extend the FEM analysis further to take into account deformation of a saturated, porous, fractured solid as well as the flow of a compressible fluid in the connected voids of pores and cracks. This extension of FEM is a coupled hydro-mechanical model.

    Reply:

    The divergence theorem is where to start as in previous derivations. Two applications are necessary. The first pertains to the virtual work of forces applied to the combined solid and fluid material; the second relates to virtual power of fluid motion relative to the solid. Symbolically,

$$\int_S \{u\}^t \{T\} dS + \int_V \{u\}^t \{\gamma\} dV = ? \tag{1a}$$

$$\int_S \{\dot{u}\}^t \{T\} dS + \int_V \{\dot{u}\}^t \{\gamma\} dV = ? \tag{1b}$$

where the right-hand sides are yet to be specified. The immediate task is to identify the variables in these two integrals. To do so requires specification of the constitutive equations for the phenomena at hand. In case of porous solid, assumption of Hooke's law is a start for poroelasticity, and in case of fluid, Darcy's law allowing for compressibility is appropriate. The total stress required for equilibrium is the sum of stress transmitted through the porous solid and stress transmitted through the fluid. Thus, conceptually, $\sigma = \sigma' + p'$, where the primes signify effective stresses that are reckoned per unit total area. In terms of forces acting over total area A, $F = \sigma A = \sigma' A + p' A = F_s + F_f$ with the same sign convention for solid and fluid stress. Recall that $A = A_s + A_f$, which implies $\sigma = \sigma_s(1-n) + \sigma_f n$ (n = porosity), again with the same sign convention for all stresses.

    In matrix form, from the "Notes", $\{\sigma\} = [E]\{\varepsilon\} + \{c\}(P)$ and also $\varsigma = cP - \{c\}^t\{\varepsilon\}$. Thus, in three dimensions,

$$\begin{Bmatrix} \{\sigma\} \\ -\varsigma \end{Bmatrix} = \begin{bmatrix} [E] & \{c\} \\ \{c\}^t & -c \end{bmatrix} \begin{Bmatrix} \{\varepsilon\} \\ P \end{Bmatrix} \tag{2}$$

Returning to (1a) in finite element notation,

$$\{\delta\}^t \left(\int_S [N]^t \{T\} dS + \int_V [N]^t \{\gamma\} dV \right) = \int_V \{\varepsilon\}^t \{\sigma\} dV)$$

$$= \{\delta\}^t \left(\int_V [B]^t (\{\sigma'\} + \{c\}(P)) dV \right)$$

$$= \{\delta\}^t \left(\int_V ([B]^t [E][B]\{\delta\} + [B]^t \{c\}[N_p]\{P_n\}) dV \right)$$

Finally for (1a), the porous solid,

$$\{f\} = [k]\{\delta\} + [k']\{P_n\} \tag{3}$$

For (1b),

$$\int_S \{\dot{u}\}^t \{T\} dS + \int_V \{\dot{u}\}^t \{\gamma\} dV = \int_S P\{n\}^t \{v\} dS + \int_V \{v\}^t \{\gamma\} dV$$

$$= \int_S P\{v_n\} dS + \int_V \{v\}^t \{\gamma\} dV$$

$$= \int_V \left(P(\partial v/\partial x + \partial v/\partial y + \partial v/\partial z) + (v_x \partial P/\partial x) \right.$$

$$\left. + (v_y \partial P/\partial y + v_z \partial P/\partial z) \right) dV + \int_V \{v\}^t \{\gamma\} dV$$

After collecting terms,

$$\int_S P\{v_n\} dS = \int_V \left(P(\partial v/\partial x + \partial v/\partial y + \partial v/\partial z) + (v_x \partial P/\partial x) + (v_y \partial P/\partial y + v_z \partial P/\partial z) \right) dV$$

In finite element form in part

$$\{P_n\}^t \int_S [N_p]^t (q) dS = \{P_n\}^t \int_V [N_p] \left(\partial(\varepsilon_v^f - \varepsilon_v^s)/\partial t + [B_p]^t \{v\} \right) dV$$

Recall that $(\partial v/\partial x + \partial v/\partial y + \partial v/\partial z)$ is the fluid volume strain rate relative to the solid volume strain rate (velocity is relative). Continuing

$$\{P_n\}^t \int_S [N_p]^t (q) dS$$

$$= \{P_n\}^t \int_V [N_p](c\dot{P} - \{c\}^t \{[B]\{\dot{\delta}\} + [B_p]^t [k]([B_p]\{P_n\} + \{\gamma_f\})) dV \text{ and}$$

$$\int_S [N_p]^t (q) dS + \int_V [B_p]^t [k]\{\gamma_f\} dV$$

$$= \int_V [N_p](c\dot{P} - \{c\}^t \{[B]\{\dot{\delta}\} + [B_p]^t [k]([B_p]\{P_n\}) dV \text{ so that}$$

$$\int_S [N_p]^t(q)\,dS + \int_V [B_p]^t[k]\{\gamma_f\}\,dV$$

$$= [k'_{pp}]\{\dot{P}\} - [k_{p\delta}]\{\dot{\delta}\} + [k_{pp}]\{P_n\}.\text{ Hence}$$

$\{f_p\} = [k'_{pp}]\{\dot{P}\} - [k_{p\delta}]\{\dot{\delta}\} + [k_{pp}]\{P_n\}$ for the
fluid portion of the problem.

For the entire hydro-mechanical formulation,

$$\{f\} = [k_{\delta\delta}]\{\delta\} + [k_{\delta p}]\{P_n\}$$

$\{f_p\} = [k'_{pp}]\{\dot{P}\} - [k_{p\delta}]\{\dot{\delta}\} + [k_{pp}]\{P_n\}$. After time integration,

$$\{f\} = [k_{\delta\delta}]\{\delta\} + [k_{\delta p}]\{P_n\} \text{ and now}$$

$\{f_p\} - \int_{\Delta t}(\{f_p\} - [k_{pp}]\{P_n\})\,dt = [k_{p\delta}]\{\delta\} - [k'_{pp}]\{P_n\}$. In compact notation

$$\begin{Bmatrix}\{f\}\\\{f_p\}\end{Bmatrix} = \begin{bmatrix}[k_{\delta\delta}] & [k_{\delta p}]\\ [k_{p\delta}] & -[k'_{pp}]\end{bmatrix}\begin{Bmatrix}\{\delta\}\\\{P_n\}\end{Bmatrix} \text{ or}$$

$$\{F\} = [K]\{\Delta\}$$

The last two forms have the advantage of symmetry. The "displacement" at a node is now three material displacements u, v, and w and a node pressure P. Each node force also has four parts; the first three are conventional forces, while the fourth part is a node velocity.

2. Program verification involves comparison of program results with known problem solutions as always. One such solution in case of deformation of saturated porous media is the famous Terzaghi consolidation problem. Solve a Terzaghi-like consolidation problem. The problem is a one-dimensional application of load to a column of a saturated porous material that otherwise follows Hooke's law in the realm of linear elasticity; fluid flow in the porous material is described by Darcy's law.

    For the porous solid, use a Young's modulus of 68.9 MPa (10,000 psi) and a Poisson's ratio of zero. The fluid is incompressible, so $c = 0$. Consider the column height $L = 100$. Hydraulic conductivity is $4.88\,(10^{-5})$ m/s [$1.602\,(10^{-4}\,\text{ft/s})$], and the specific weight of water is $9.88\,\text{kN/m}^3$ ($62.43\,\text{lbf/ft}^3$).

    An example plot of some pressure results is given in Figure 1. Be sure to find displacement as a function of time and depth as well.

    Reply:
    Consider the schematic in the sketch.
    Sketch for a consolidation problem.

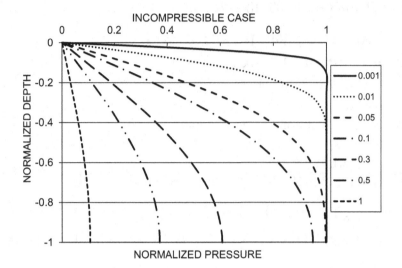

Figure 1 Pressure in the consolidating column at various times. Normalized depth is actual depth divided by column depth. Normalized pressure is actual pressure
divided by the applied pressure. Times in the legend are dimensionless times
obtained with the aid of the consolidation coefficient that is briefly discussed
below in notes.

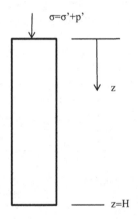

Basics:

Sides and bottom are impermeable, no flow.
Flow occurs out of the top under total stress σ.
Sides and bottom are also fixed in the perpendicular directions.

Equilibrium of total stress:

$$\partial\sigma/\partial z + \gamma = 0$$

Equilibrium of fluid stress:

$$\partial p'/\partial z + \gamma_f = R$$

where $p' = \pi$ in "Notes".

Because no lateral displacement is allowed, horizontal strains in the porous, saturated solid are zero. The vertical strain is then according to Hooke's law (isotropic, linearly elastic solid)

$$\varepsilon_{zz} = \sigma'_{zz}(1+v)(1-2v)/E(1-v)$$

where the primes denote effective stress (total stress for equilibrium; effective stress for material behavior).

Because the horizontal strains are zero, the volumetric strain in the solid is

$$\varepsilon_v^s = \sigma'_{zz}/E' = (\sigma_{zz} - p')/E', \quad E' = \frac{E(1-v)}{(1+v)(1-2v)}$$

where use is made of the definition of effective stress.

The volumetric strain of the fluid, when considered compressible, may be expressed as

$$\varepsilon_v^f = cp'$$

where c is fluid compressibility. In the case of incompressibility, $c = 0$.

Darcy's law may be stated as

$$\partial w/\partial t = k\left(\partial p'/\partial z + \gamma_f\right)$$

where w is the relative displacement of fluid with respect to the solid; the time derivative is the Darcy velocity.

The derivative of w with respect to z is a relative strain and in this case is the volumetric strain of the fluid relative to the solid. Thus,

$$\partial w/\partial z = \varepsilon_v^f - \varepsilon_v^s = cp' - (\sigma_{zz} - p')/E'$$

The time derivative of this last expression is

$$\partial^2 w/\partial z\,\partial t = (c + 1/E')\partial p'/\partial t$$

The same mixed derivative can be obtained from Darcy's law. Thus,

$$\partial^2 w/\partial z\,\partial t = k\partial^2 p'/\partial z^2$$

Equating the two mixed derivatives leads to

$$\underline{\partial p'/\partial t = c_v\,\partial^2 p'/\partial z^2}, \text{a diffusion type of differential equation,} \tag{1}$$

where c_v is known as the "consolidation coefficient". In fact, $c_v = kE'/(1 + E')$, and again, in case of incompressible fluid, $c = 0$. If the solid were incompressible, no motion would occur and consolidation would not be possible.

Solution to Equation (1) may be obtained by a standard variable separable technique. Thus, assuming $p' = T(t) Z(z)$, after differentiation and substitution into (1), one obtains

$$T'/c_v T = Z''/Z = -\lambda^2$$

where the right-hand side is a constant. Hence,

$$T' + \lambda^2 c_v T = 0 \quad \text{and} \quad Z'' + \lambda^2 Z = 0$$

Solutions to these two ordinary differential equations may be obtained in the usual way. Thus,

$$T = Ce^{-c_v\lambda^2 t} \quad \text{and} \quad Z(z) = A\cos(\lambda z) + B\sin(\lambda z)$$

where A, B, and C are integration constants to be determined from boundary and initial conditions as always. The result is then

$$p' = [a\cos(\lambda z) + b\sin(\lambda z)][\exp(-c_v\lambda^2 t)] \tag{2}$$

where the order and notation have been changed for ease of reading that products of constants lead to new constants a and b.

The pressure at the top of the column is open to the atmosphere and is thus zero. However, the pressure inside the column at the instant of load application is equal to the load because the solid has no time to deform, so the entire load is carried by the fluid. Thus,

(i) $p'(0, z) = \sigma = p_o$ and (ii) $p'(t, 0) = 0$.

Applying the first condition (i) to (2) implies $a = 0$, so the cosine term drops out.

At the bottom of the column, no flow occurs, so the pressure gradient must be 0 in view of Darcy's law, that is,

(iii) $\partial p'/\partial z(t, H) = b\lambda^2 \cos(\lambda H)\exp(-c_v\lambda^2 t)$.

Hence, $\lambda H = (2n - 1)\pi$, $n = 1, 2, 3,\ldots$. For every n, one obtains a λ and yet another solution, that is, another sine term that requires another constant b. Indeed, the solution is a Fourier sine series and the constant terms are determined in the usual way: multiply by another sine term with a different λ, observe the orthogonal nature of the product of the sine terms except when they have the same λ, then integrate with respect to z from 0 to H, etc. The result is

(iv) $b_n = 4p_o/\pi(2n - 1)$ or $b_n = 4p_o/\pi m$, $m = 1, 3, 5,\ldots$

Finally,

$$p' = p_o\left[\sum_{m=1,3,\ldots} (4/m\pi)\sin(m\pi z/2H)\left[\exp(-c_v m^2\pi^2 t/4H^2)\right]\right] \tag{3}$$

which is Equation (11.12) in the Notes. Figure 11.5 is a schematic plot of stress, effective stress, and pressure as time passes (incompressible fluid case). If the fluid were compressible, there would be an instantaneous drop of the porous piston through which the load is applied.

Use of (3) and Hooke's law leads to an expression for strain that can be integrated to obtain displacement as a function of time and depth. Of course, displacement of most interest is displacement of the surface where the load is applied ($z = 0$). After integration, the result is

$$w(z, t) = \left(\frac{p_o H}{E} \right) \left[\sum_{m=1,3,...}^{m=\infty} \left(\frac{8}{m^2 \pi^2} \right) \cos\left(\frac{m \pi z}{2H} \right) \left(1 - \exp\left[-\left(\frac{m^2 \pi^2}{4H^2} \right) c_v t \right] \right) \right] \tag{4}$$

Inspection of (4) shows that the initial displacement everywhere in the column is zero at $t = 0$ and when the surface load is just applied. If the fluid were compressible, an instantaneous elastic displacement would occur upon application of load. The surface displacement is clearly

$$w(0, t) = \left(\frac{p_o H}{E} \right) \left[\sum_{m=1,3,...}^{m=\infty} \left(\frac{8}{m^2 \pi^2} \right) (1) \left(1 - \exp\left[-\left(\frac{m^2 \pi^2}{4H^2} \right) c_v t \right] \right) \right] \tag{5}$$

Figure 2 shows surface displacement as a function of time according to (5), which was evaluated using 100 terms.

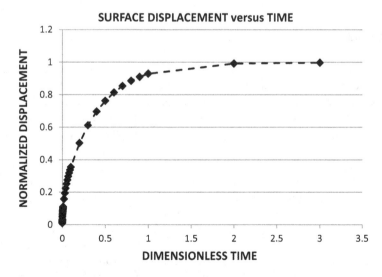

Figure 2 Surface displacement as a function of time, incompressible fluid case. Displacement is normalized by division by long time displacement. Dimensionless time is obtained through the consolidation coefficient.

An interesting "trick" to obtain the long-term steady-state equilibrium condition is to use a large time step. Examination of the analytical solution shows this feature to be the case, provided one is cognizant of the series solution and the number to which it converges.

Notes: Units of the consolidation coefficient c_v are (L^2/T) as inspection of the diffusion equation shows. Our computation expressed $c_v = \dfrac{kE'}{1+cE'}$. The question that arises concerns the units of the "coefficient of permeability". The answer depends on how Darcy's law is written. When written using pressure gradients, the units of k are (L^4/FT). A more convenient form is the use of head gradients, so k is hydraulic conductivity with units of velocity (L/T). Hence, $c_v = \dfrac{(k/\gamma_f)E'}{1+cE'}$, which has units of $c_v = \dfrac{\left[(L/T)/(F/L^3)\right](F/L^2)}{1+[1/(F/L^2)](F/L^2)} = (L^2/T)$. Units are important!

An important use of c_v is in the formation of a dimensionless time t^* such that $t^* = tc_v/L^2$ and $t = t^* L^2/c_v$. A characteristic time for consolidation to be almost complete is often approximated by a dimensionless time of one, so the actual time in a given problem is $t = L^2/c_v$. Time steps starting small at 0.0001 or so and increasing by a factor of 10 are often used in consolidation studies. Of course, the units of time in practice may be seconds, days, or years, depending on the material and the problem.

Another interesting and practical consideration besides bearing capacity of foundations occurs during blasting in wet ground. This action may be considered as applying a sudden upward force instead of a downward one in the Terzaghi consolidation. What happens to the ground about the blast or to the solid and fluid stresses in the column?

GEOFEM 13 KEY

1. Verify a hydro-mechanical couple finite element program by comparing FE results with the theoretical solution to the consolidation problem addressed previously.

   Generate a simple mesh for validation purposes, say, with a 14-element column of two-dimensional elements with nine equal square elements from the bottom ($z = +$up) and with partitioning of the tenth element at the column top into five thin elements. This refinement at the top of the mesh gives more details where force is applied.

   In this regard, there is a pressure discontinuity at the surface that usually shows as fluctuations in the first few top elements and in the Fourier series solution at small time steps, especially so if the fluid is considered incompressible. The solid is compressible in any case, or else no deformation is possible. When both are compressible, the surface undergoes an instantaneous elastic displacement as the applied total load is distributed between fluid and solid. In case of incompressible fluids and very small time steps (near zero), the applied load is carried entirely by the fluid. The pressure gradient is high at this instant, so the instantaneous fluid velocity is also very high.

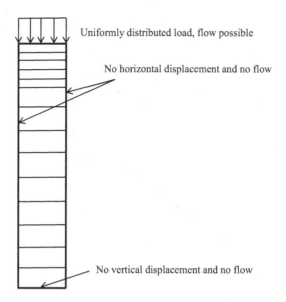

Uniformly distributed load, flow possible

No horizontal displacement and no flow

No vertical displacement and no flow

Figure 1 A 14-element mesh for coupled problem analysis.

Comparison of stresses (total, effective, and fluid) via plots of data is perhaps the best way of doing code validation. However, displacements could also be compared, provided a displacement solution is computed from the fluid stress solution. Quantitative comparison of results is readily done through regression analysis.

2. Continue with material properties used in the previous exercise. Hence, for the porous solid, use a Young's modulus of 68.9 MPa (10,000 psi) and a Poisson's ratio of zero. The fluid is incompressible, so $c = 0$. Consider the column height $L = 100$. Hydraulic conductivity is 4.88 (10^{-5}) m/s [1.602(10^{-4}ft/s)], and the specific weight of water is 9.88 kN/m^3 (62.43 lbf/ft^3).

Reply:

Pressure results using a two-dimensional coupled hydro-mechanical finite element program are shown in Figure 2. Agreement of FEM results with theory is excellent as the regression line equation and correlation coefficient show. Results are for the incompressible fluid case as requested. Normalization is achieved by dividing current pressure with applied pressure, so final normalized pressures are unity as seen in the figure.

Figure 3 shows regression of finite element surface displacement on theoretical surface displacement. Agreement is excellent as seen in the regression line equation.

The coupled finite element program used in these analyses is a two-dimensional program that uses a 4CST element for these poroelastic analyses. The theoretical analysis is through Fourier series solution of the consolidation problem posed for this study. The program used allows for a choice of "theta" in time integration (forward, backward, and Crank–Nicolson). Also allowed is deformation beyond the elastic range. When doing poroelastic–plastic analysis, incremental loading is necessary for iterative solution. In such cases, the load

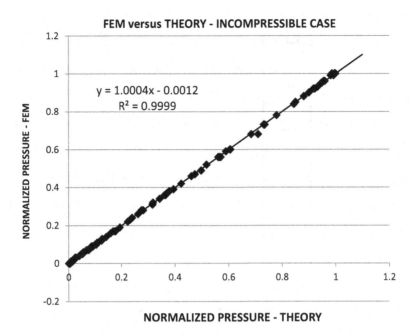

Figure 2 Regression of finite element pressure on theoretical pressure in normalized or dimensionless form. There are 98 data points obtained in seven time steps that are used in the regression analysis.

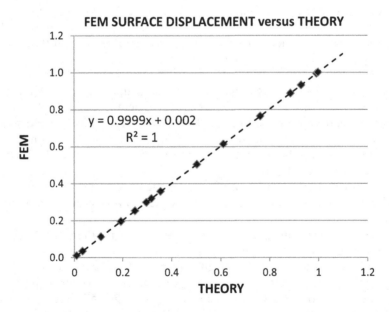

Figure 3 Regression of finite element surface displacement on theoretical displacement. Displacements are normalized by the displacement in the long term.

increment is conveniently synchronized with the time increment. The poroelastic–plastic case poses a moving boundary problem (the elastic–plastic boundary) that is similar to a freezing front problem. This problem defies series solution, but is still amenable to analytical solution but only for a column of infinite depth.

Be sure to mention the equation of solving technique and the type of element used in your finite element program and to comment on agreement or lack with theory.

Reply:

An elimination equation solver was used in this quite small finite element problem.

Note: An important use of c_v is in the formation of a dimensionless time t^* such that $t^* = tc_v/L^2$ and $t = t^* L^2/c_v$. A characteristic time for consolidation to be almost complete is often approximated by a dimensionless time of one, so the actual time in a given problem is $t = L^2/c_v$. Time steps starting small at 0.0001 or so and increasing by a factor of 10 are often used in consolidation studies. Of course, the units of time in practice may be seconds, days, or years, depending on the material and the problem.

Another interesting and practical consideration besides bearing capacity of foundations occurs during blasting in wet ground. This action may be considered as applying a sudden upward force instead of a downward one in the Terzaghi consolidation. What happens to the ground about the blast or to the solid and fluid stresses in the column?

GEOFEM 14 KEY

Use a boundary element method (BEM) to compute the stress distribution about twin circular tunnels excavated in a stress field such that the vertical stress is 10 MPa and the horizontal stress is 0 MPa. Young's modulus is 10 GPa, and Poisson's ratio is 0.25. Separation of tunnels is equal to tunnel diameter.

1. Determine the stress concentration factors at points A, B, C, and D ("Notes", Figure 13.4) assuming the applied stress (preexcavation) is compressive; that is, compression is positive. Also determine the vertical stress at the center of the pillar that separates the twin tunnels. Hint: Vertical stress applies at A, B, and D; horizontal stress applies at C. Stress concentration is the actual stress divided by the major principal stress applied (10 MPa). Tabulate results.
   Reply: BEM results are entered in the table.
2. Repeat the analysis with a horizontal loading of 10 MPa. Tabulate results.
   Reply: BEM results are entered in the table.
3. Estimate the stress concentrations at A, B, C, and D from the table values when the preexcavation stresses are $SV = 10$, $SH = 5$ and then compute using BEM. Tabulate results.
   Reply: BEM results are entered in the table.
4. Repeat analyses using FEM and then enter results into the table with BEM results.
   Reply: FEM results are entered in the table.
   Table of Stress Concentrations at Critical Points at the Walls of Twin Circular Tunnels.

Point/load	Method	A	B	C	D
SV = 10, SH = 0	BEM	3.06	2.77	−0.59	1.48
	FEM	2.75	2.73	−0.70	1.41
SV = 0, SH = 10	BEM	−0.90	−0.48	2.36	0.05
	FEM	−0.73	−0.49	2.45	0.07
SV = 10, SH = 5 EST	BEM	2.21	2.47	0.59	1.51
	FEM	2.39	2.48	0.53	1.45
SV = 10, SH = 5 COMP	BEM	2.56	2.53	0.59	1.51
	FEM	2.37	2.48	0.56	1.44

5. *Reply*:
   There are noticeable differences between BEM and FEM. This result should not be surprising because BEM results are at excavation boundaries or very near. FEM results are at element centers that are necessarily away from the excavation boundaries. BEM results are sensitive to point picks near excavation boundaries, while FEM results depend on mesh refinement and are always away from boundaries. In application of both methods, the accuracy is certainly acceptable for engineering design in geomechanics where spatial variability of material properties is more the rule than the exception.

GEOFEM 15 KEY

This exercise concerns the distinct element method (DEM) and discontinuous deformation analysis (DDA).

1. Diagram in schematic form the first of Equations (15.2) in the text.
   Reply: With reference to the diagram below where velocity is on the y-axis and time t is on the x-axis, the difference between velocities at time $t - \Delta t/2$ and $t + \Delta t/2$ is $\dot{u}(t + \Delta t/2) - \dot{u}(t - \Delta t/2)$ and thus the acceleration $d\dot{u}/dt$ is approximately

$$d\dot{u}/dt = \left[\dot{u}(t - \Delta t/2) - \dot{u}(t + \Delta t/2)\right]/\Delta t$$

which is the first of 15.2 in the text.

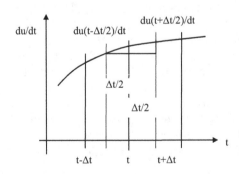

2. Justify Equations (15.4) in the text with the aid of a diagram. Focus on the first of 15.4.

   Reply: With reference to the diagram below

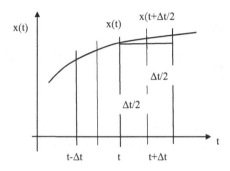

$$du/dt = \left[x(t + \Delta t/2) - x(t) \right] / \left[\Delta t/2 \right].$$

Solving for $x(t + \Delta t/2)$ gives

$$
\begin{aligned}
x(t + \Delta t/2) &= x(t) + [du/dt](\Delta t/2) \\
&= x(t) + [u(t + \Delta t) - u(t)](\Delta t/2) \\
&= x(t) + 2[\dot{u}(t + \Delta t/2)](\Delta t/2) \\
x(t + \Delta t/2) &= x(t) + \dot{u}(t + \Delta t/2)\Delta t
\end{aligned}
$$

where $[u(t + \Delta t) - u(t)] = 2[\dot{u}(t + \Delta t/2)]$ is used.

3. Show that without damping, DEM block collisions will continue indefinitely. Do this demonstration by dropping a single ball or disk of specified mass from a given height. Solve the problem of energy conservation and then use a DEM program such as BALL or TRUBAL to obtain results for comparison with the exact solution.

   Reply:

   Consider a ball dropped in the gravity field and subsequent elastic rebound after striking a horizontal base. Elastic rebound implies a coefficient of restitution of one. Conservation of energy dictates rebound to the original height after the fall. When the ball reaches the original height, velocity reverses and the process is repeated.

   Conservation of energy in this case is simply $\Delta E = \Delta KE + \Delta PE = 0$, where $\Delta E, \Delta KE$, and ΔPE are changes in energy, kinetic energy, and potential energy in the gravity field. Work and heat transfer are zero. Thus, $mv^2/2 = mgh$, where m, v, g, and h are mass, velocity, acceleration of gravity, and height of fall. (Note: The ball starts from rest.) Hence, the velocity at the end of the fall is $v = \sqrt{2gh}$ and the time of fall is $t = \sqrt{2h/g}$.

   A DEM demonstration is given by Acharya (1991)[1] and is illustrated in the following figure sequence that shows ball positions at various time intervals. Theoretical distance of free fall is 0.7666 m that occurs in 0.3954 s. Fall distance in DEM is 0.7619 m. The relative error is 0.6%.

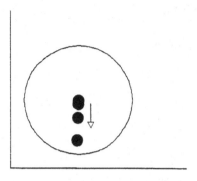
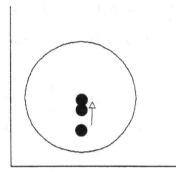

Single ball dropping 0.0000–0.3954 s. Single ball rebounding 0.5272–0.7908 s.
Note: Round trip travel time is (2) (0.3954) = 0.7908 s.

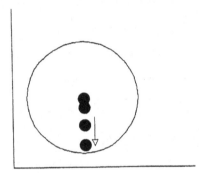
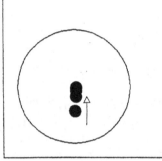

Single ball dropping 0.9226–1.3180 s. Single ball rebounding 1.4498–1.8452 s.
Note: Travel time down and travel time up are 0.3954 s.

4. Block motion is composed of displacements associated with translation and rotation of the mass center and volume and shape changes. Show in two dimensions
(x, y) that displacements (u, v) in the x- and y-directions are given by

$$u = u_o - (y - y_o)\omega + (x - x_o)\varepsilon_{xx} + (y - y_o)\gamma_{xy} \qquad (1)$$

where u, u_o, ω, ε_{xx}, and γ_{xy} are displacement, translation displacement, rotation
angle (radians) normal strain $(\partial u / \partial x)$, and shear strain $(1/2) (\partial u / \partial y + \partial v / \partial x)$.
Similarly for v,

$$v = v_o + (x - x_o)\omega + (y - y_o)\varepsilon_{yy} + (x - x_o)\gamma_{xy} \qquad (2)$$

with normal strain $\varepsilon_{yy} = (\partial v / \partial y)$.
How large may the rotation be and can it still be considered small?
Reply: Rigid body translation of the mass center has displacement u_o in the
x-direction, and rigid body displacements from rotation about the mass center are
given by the cross product (determinant)

$$\begin{vmatrix} i & j & k \\ 0 & 0 & 0 \\ (x-x_o) & (y-y_o) & 0 \end{vmatrix}, \text{ where } i, j, k \text{ are unit vectors along the } x\text{-}, y\text{-},$$

z-axes.

Hence,

$-(y-y_o)\omega$, $(x-x_o)\omega$, 0 are displacements in the x-, y-, z-directions, respectively. These displacements are common to all points in the considered block.

Displacements associated with normal strains are approximately

$$u = (x-x_o)\varepsilon_{xx}, \quad v = (y-y_o)\varepsilon_{yy}.$$

With reference to the diagram below, shear strain γ is the angular change indicated and for small strains, the tangent of the angle is closely approximated by the angle as such, that is, $\tan(\gamma) \doteq \gamma$.

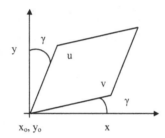

Hence, $\tan(\gamma) = \gamma = v/(x-x_o) = u/(y-y_o)$ allows for determination of displacements u and v associated with shear strain. Here $\gamma = \gamma_{xy}$.

Addition of displacements gives Equations (1) and (2) for a *forward* analysis.

Alternatively, one may begin with the interpolation function for a linear strain triangle

$$u = a_o + a_1 x + a_2 y$$

Then consider a rigid body translation of the center (and all other points in the triangle) such that

$u_o = a_o + a_1 x_0 + a_2 y_0$. Hence,

$u = u_o + a_1(x-x_o) + a_2(y-y_o)$ and

$v = v_o + b_1(x-x_o) + b_2(y-y_o)$. Thus,

$\partial u/\partial x = a_1, \partial u/\partial y = a_2, \partial v/\partial x = b_1, \partial v/\partial y = b_2$. Therefore,

$a_1 = \varepsilon_{xx}, b_2 = \varepsilon_{yy}, (1/2)(a_2+b_1) = \gamma_{xy}, (1/2)(b_1-a_2) = \omega$. After substitution,

$u = u_o + (x-x_o)\varepsilon_{xx} + \gamma_{xy}(y-y_o) - w(y-y_o)$
$v = v_o + (x-x_o)\gamma_{xy} + \varepsilon_{yy}(y-y_o) + w(x-x_o)$ which are Equations (1) and (2).

The rotation angle is limited by the approximation $\tan(\gamma) = \gamma = v/(x-x_o) = u/(y-y_o)$. A tabulation of $\tan(\gamma)$ and γ provides guidance:

γ (radians)/(degrees)	$\tan(\gamma)$	Relative error %
0/0	0	0
0.698/4	0.699	0.1
0.1396/8	0.1405	0.6
0.2094/12	0.2126	1.5
0.2793/16	0.2867	2.6
0.3491/20	0.3640	4.1
0.4189/24	0.4452	5.9

Table results indicate that a rotation angle of less than 20° incurs an error less than 4.1%.

5. Show that a least square fit of displacements at three measurement points in a two-dimensional block allows for interpolation of displacements elsewhere in the block similar to finite element analysis. Consider the three measurement points in a triangular configuration as illustrated in the diagram following.

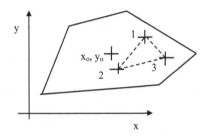

Reply: A block with three points, 1, 2, 3, forming a triangle is shown in the sketch. A least square function for minimization is $\Phi = \sum\limits_{i=1}^{i=3}\left[(u_i - \bar{u}_i)^2 + (v_i - \bar{v}_i)^2\right]$, where the overbar indicates a measured value.

For a stationary value, the first derivatives must vanish. Hence, $\partial\Phi/\partial u_i = 2(u_i - \bar{u}_i) = 0$ and $\partial\Phi/\partial v_i = 2(v_i - \bar{v}_i) = 0$, which are clearly satisfied by equating the unknown displacements to the measured displacements.

A linear interpolation scheme has the familiar finite element form $u = a_o + a_1 x + a_2 y$ and $v = b_o + b_1 x + b_2 y$. In view of the known displacements, one has $\begin{aligned}\bar{u}_1 &= a_o + a_1 x_1 + a_2 y_1\\ \bar{u}_2 &= a_o + a_1 x_2 + a_2 y_2\\ \bar{u}_3 &= a_o + a_1 x_3 + a_2 y_3\end{aligned}$ that allows for solution of the three unknown constants, a_o, a_1, a_2, and similarly for b_o, b_1, b_2. After solution for the constants and back-substitution as in the finite element analysis for a linear displacement triangle, one has the same form for interpolation. Thus, for the block

$$\begin{Bmatrix}u\\v\end{Bmatrix} = \begin{bmatrix} N_1 & 0 & N_2 & 0 & N_3 & 0 \\ 0 & N_1 & 0 & N_2 & 0 & N_3 \end{bmatrix}\begin{Bmatrix}\bar{u}_1\\\bar{v}_1\\\bar{u}_2\\\bar{v}_2\\\bar{u}_3\\\bar{v}_3\end{Bmatrix} \qquad (3)$$

where the Ns are the interpolation functions of x and y found in finite element analysis of a linear displacement triangle or from the analysis in this problem. Equation (3) follows from a *back-analysis* involving measurements of block displacement.

6. Consider a triangular block in two dimensions (x, y) with corners 1, 2, 3 at $(2, 2)$, $(4, 1)$, and $(3, 3)$, respectively. Measured displacements (u, v) are given in the table following.

Table of displacements for Problem 6.

	u	v
1	1.515192	1.326352
2	1.658456	1.68884
3	1.326352	1.484808

Sketch the original and final positions of the block and then determine block displacements associated with translation, rotation, and strains.

$$u = a_o + a_1 x + a_2 y$$
$$v = b_o + b_1 x + b_2 y \tag{1}$$

Reply: The sketch required is shown below and illustrates the original position, rotated position, and translated and rotated position as well. A line connects the centers of the original and final positions.

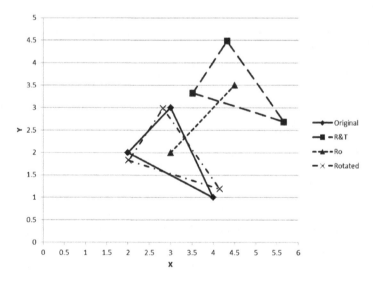

Sketch for Problem 6 showing original triangular block and final position.

$$\text{The interpolation scheme leads to the equation} \begin{Bmatrix} u_1 \\ u_2 \\ u_3 \end{Bmatrix} = \begin{bmatrix} 1 & x_1 & y_1 \\ 1 & x_2 & y_2 \\ 1 & x_3 & y_3 \end{bmatrix} \begin{Bmatrix} a_o \\ a_1 \\ a_2 \end{Bmatrix}$$

that allows for solution of the constants, the as and similarly for the bs.

a_o	1.892873096	b_o	1.009439961
a_1	-0.015192247	b_1	0.173648178
a_2	-0.173648178	b_2	-0.015192247

Translation of the center (3, 2) is then given by

$u_o = 1.8929 + (-0.0152)(3) + (-0.1736)(2)$
$v_o = 1.0094 + (0.1736)(3) + (-0.0152)(2)$. Thus,

$u_o = 1.5$
$v_o = 1.5$ are the rigid body displacements of the triangle.

In general, after subtracting the rigid body translation from the displacements (u, v), one obtains

$u = u_o + a_1(x - x_o) + a_2(y - y_o)$
$v = v_o + b_1(x - x_o) + b_2(y - y_o)$ where $x_o = 3.0$, $y_o = 2.0$.

The rigid body rotation by definition is $\omega = (1/2)(\partial v/\partial x - \partial u/\partial y)$, that is, $\omega = (1/2)(b_1 - a_2)$. Normal strains are $\varepsilon_{xx} = \partial u/\partial x$, $\varepsilon_{yy} = \partial v/\partial y$, and shear strain is $\gamma_x = (1/2)(\partial u/\partial y + \partial v/\partial x)$. Thus, $\varepsilon_{xx} = a_1$, $\varepsilon_{yy} = b_2$, and $\gamma_x = (1/2)(a_2 + b_1)$. After substituting these relations into the equations for displacement, one obtains

$u = u_o - (y - y_o)\omega + (x - x_o)\varepsilon_{xx} + (y - y_o)\gamma_{xy}$
$v = v_o + (x - x_o)\omega + (y - y_o)\varepsilon_{yy} + (x - x_o)\gamma_{xy}$ (2)

The rigid body rotation of the triangular block is

$\omega = (1/2)(b_1 - a_2)$
$= (1/2)(0.17365 - (-0.17365))$
$\omega = 0.17365$ radians, $10°$.

This rotation is "small" in the sense that the angle tangent and angle are approximately equal.

Interestingly, the normal strains are not zero by this analysis, although the problem set up is based on zero strains. Specifically,

$\varepsilon_{xx} = a_1$ $\varepsilon_{xx} = -0.0152$
$\varepsilon_{yy} = b_2$, that is, $\varepsilon_{yy} = -0.0152$
$\gamma_{xy} = (1/2)(a_2 + b_1)$ $\gamma_{xy} = (1/2)(-0.1736 + 0.1736)$

which shows non-zero normal strains and zero shear strain. The reason is not clear, but perhaps the large displacements are incompatible with small strain.

NOTES

1 Acharya, A. (1991) "A Discrete Element Approach to Ball Mill Mechanics." M.S. Thesis, University of Utah, p 103.

Index

Note: **Bold** page numbers refer to tables; *italic* page numbers refer to figures and page numbers followed by "n" denote endnotes.